T0134565

# Signals and Communication Technology

More information about this series at http://www.springer.com/series/4748

Mohsen A. M. El-Bendary

# Wireless Personal Communications

## Simulation and Complexity

Mohsen A. M. El-Bendary
Department of Electronics Technology
Helwan University
Cairo
Egypt

ISSN 1860-4862 ISSN 1860-4870 (electronic)
Signals and Communication Technology
ISBN 978-981-13-3917-2 ISBN 978-981-10-7131-7 (eBook)
https://doi.org/10.1007/978-981-10-7131-7



Printed on acid-free paper

This Springer imprint is published by the registered company Springer Nature Singapore Pte Ltd.
part of Springer Nature
The registered company address is: 152 Beach Road, #21-01/04 Gateway East, Singapore 189721, Singapore

***To my mother and father soul and family***

*I dedicate this book to my parents, my wife, and my children who supported me all the time with their love and pray.*

Mohsen A. M. El-Bendary

# Preface

This book introduces wireless personal communications from the point of view of wireless communication system researchers. Existing sources on wireless communications put more emphasis on simulation and fundamental principles of how to build a study model. In this volume, the aim is to pass on to readers as much knowledge as is essential for completing model building of wireless communications, focusing on wireless personal area networks (WPANs).

This book is the first of its kind that gives step-by-step details on how to build the WPANs simulation model. It is most helpful for readers to get a clear picture of the whole wireless simulation model by being presented with many study models. The book is also the first treatise on wireless communication that gives a comprehensive introduction to data-length complexity and the computational complexity of the processed data and the error control schemes.

This volume is useful for all academic and technical staff in the fields of telecommunications and wireless communications, as it presents many scenarios for enhancing techniques for weak error control performance and other scenarios for complexity reduction of the wireless data and image transmission. Many examples are given to help readers to understand the material covered in the book. Additional resources such as the MATLAB codes for some of the examples also are presented.

It proposes many scenarios for data and image wireless communications over fixed and mobile WPANs, focusing on the complexity of each one.

It presents numerous examples to help readers understand the material covered in the book and provides additional resources such as the MATLAB codes for some of the examples in the book.

This book presents some methods to simulate and study the WPAN-Bluetooth network. Also, it presents Matlab codes of different proposed scenarios to improve the WPAN-Bluetooth system performance through channel coding schemes. It presents an analysis of the Bluetooth system performance over different communication channels with standard Bluetooth channel coding scheme. The book discusses the effect of the channel coding on Bluetooth system performance. It proposes other block coding schemes and studies the difference in Bluetooth system performance in the presence of these schemes.

It presents modified Bluetooth frame format by using convolutional codes with different Bluetooth frame portions. Book studies the performance of Bluetooth system in case of proposed error control codes. Also, this book shows the effects of longer K (constraint length) on basic Bluetooth ACL packets' performance over AWGN and Rayleigh-flat fading channels.

This book proposes the use of pseudo coding based on chaotic maps in the Bluetooth communication systems for data encryption prior to transmission. The purpose of this step is to reduce the effects of fading on the transmitted signals. Also, this chapter studies the effect of block coding on the performance of uncoded packet.

Cairo, Egypt Mohsen A. M. El-Bendary

# Contents

# Abbreviations and Symbols

| | |
|---|---|
| AC | Access code |
| ACL | Asynchronous connectionless |
| ARQ | Automatic repeat request |
| AWGN | Additive white Gaussian noise |
| BD_ADDR | Bluetooth device address |
| BER | Bit error rate |
| BGFSK | Binary Gaussian Frequency Shift Keying |
| BNEP | Bluetooth network emulation protocol |
| CAC | Channel access code |
| DAC | Device access code |
| DHx | Data high rate |
| DMP | Device manager protocol |
| DMx | Data medium rate |
| EDR | Enhanced data rate |
| FEC | Forward error correction |
| FHS | Frequency hopping spectrum |
| FHSS | Frequency hopping spread spectrum |
| HCI | Host Controller Interface |
| HD | Header of packet |
| $HV_x$ | High-quality voice |
| IAC | Inquiry access code |
| IrDA | Infrared data association |
| ISM | Industrial, scientific, and medical band |
| k | Input of encoder |
| K | Constraint length |
| L2CAP | Logic link and adaptation protocol |
| LAN | Local area network |
| LOS | Line of sight |
| LAP | Lower Address part |
| LC | Link controller |

| | |
|---|---|
| LMP | Link manager protocol |
| LT_ADDR | Logical transport address |
| m | Redundant bits |
| n | Output of encoder |
| OBEX | OBject EXchange |
| PDA | Personal device assistant |
| PDU | Protocol data unit |
| PEP | Packets error probability |
| PER | Packet of error rate |
| PL | Payloads of packets |
| PSK | Phase Shift Keying |
| QoS | Quality of service |
| R | Code rate |
| RF | Radio frequency |
| RFCOMM | Radio frequency communication |
| RSSI | Received signal strength Indicator |
| SCO | Synchronous connection-oriented |
| SDAP | Service discovery application |
| SIG | Special interest group |
| SNR | Signal-to-Noise ratio |
| TDD | Time division duplex |
| TDMA | Time division multiple access |
| UAP | Upper address part |
| UART | Universal asynchronous receiver transmitter |
| WLAN | Wireless local area network |

# List of Figures

# List of Tables

# Chapter 1
# Introduction

## 1.1 Introduction

Over recent years, there has been a great growth in wireless technologies and its applications fields, especially in field of wireless computer network and wireless personal area networks (WPANs). The Bluetooth technology is one of WPANs. The WPANs are the personal communication technologies. The Bluetooth specifications are developed and licensed by the Bluetooth Special Interest Group (SIG) [1]. The operation range of Bluetooth systems is the unlicensed 2.4 GHz ISM band. In this band, there are many sources of interference such as WLAN (IEEE 802.11 standard).

In case of radio communications, the channel of communications systems suffers from different attacks, noise and interference which works to corrupt the transmitted signals. The number of errors at the receiver depends on the amount of noise and interference in the radio channel. Error control code schemes are essential in wireless communications to protect the transmitted information from channel effect [2].

Channel coding aims to error detection only or error detection and error correction. In case of error detection, the receiver detects errors in the received message and sends Automatic Repeat reQuest (ARQ) to the transmitter to resend the message. In the case of error detection and correction channel coding schemes, a receiver detects errors in the received message, it corrects these errors without a need for the ARQs. The main purpose of channel coding in Bluetooth systems is reducing the retransmission times and decreasing or eliminating the channel effects on the transmitted signals. There are two types of channel coding; block coding and convolutional coding. In Bluetooth systems, block coding is utilized in the Bluetooth frames encoding. Several types of block coding schemes are used in the coding of the different sections of Bluetooth frame.

The different WPANs simulation with Matlab codes are presented in this book to achieve rapid, quick and simple simulation model for studying the WPANs performance. FEC utilization in different WPANs technologies is variant, weak



M. A. M. El-Bendary, *Wireless Personal Communications*,
Signals and Communication Technology,
https://doi.org/10.1007/978-981-10-7131-7_1

FEC technique is utilized in Bluetooth standard while in ZigBee networks there no FEC is used. It is presented for near and direct to communicate different things in specific small areas with limited required power.

## 1.2 Bluetooth Definition

Bluetooth is a wireless technology that is composed of hardware, software, and operation requirements. Bluetooth is a short-range, low-power, and low-cost wireless communication technology. This technology enables its units to communicate with each other when they are in range, Bluetooth units should not have a line of sight to communicate with each other, and a line-of-sight condition is not required in Bluetooth setup connection [3]. Devices are connected through a radio channel to create WPAN.

Bluetooth has emerged as a wireless communication technology aiming at eliminating interconnection cables and connecting computer peripherals, mobile phones, and personal device assistant (PDA) using a radio link. WLANs having the IEEE 802.11b standard and Bluetooth use the same frequency range, but there are many differences between them. The IEEE 802.11 standard has no cable connection for LANs, but Bluetooth has no cable for a variety of applications. Bluetooth has the IEEE 802.15.1 standard.

## 1.3 Bluetooth Special Interest Group (SIG)

The Bluetooth Special Interest Group (SIG) has developed the Bluetooth specifications in 1998. The development of Bluetooth system standard started with the formation of the SIG by IBM, Intel, Nokia, and Toshiba in 1998. Members of the SIG worked for a global standard. In 1999, 3Com, Lucent Technologies, Microsoft, and Motorola joined the initial group as the promoter of Bluetooth Special Interest Group (SIG). In 2007, the SIG had over 9000 member companies that are leaders in communications, music, automotive, network industries, and industrial automation [4].

## 1.4 Origin of Bluetooth Name

Bluetooth is a historical word. This name was intended to be temporary because no one has introduced a better name than it. The name follows a Danish king called Harald Blatand also known as Harald Bluetooth [5]. This king is famous for his success in uniting the Scandinavian people in the tenth century AD. Also, the Bluetooth standard is supposed to unite personal devices.

Today, Bluetooth wireless technology enables electronic devices to talk to each other, but this time by means of a low-cost, short-range radio link.

## 1.5 Applications and Aspects of Bluetooth

Bluetooth is a wireless technology which has a global standard. Bluetooth has the following advantages:

1. Elimination of wires and cables between both stationary and mobile Bluetooth units.
2. Able to transfer both data and voice.
3. Supports wireless ad hoc networks. Connection of Bluetooth units is self-configured without user inputs.
4. Low cost.
5. Easy extension of a Bluetooth network.

**Bluetooth technology has the following disadvantages:**

1. It has a limited range of up to 100 m.
2. Interference with other devices is a major problem. In the Bluetooth, frequency range of operation is shared by other technologies such as baby monitors, garage door openers, cordless phones, and microwave ovens.
3. It has low data transfer rate compared to the WLAN IEEE 802.11b data rate.
4. The channel coding scheme which used to protect Bluetooth packets is not powerful.

Bluetooth applications are so wide. There are some prevalent applications of Bluetooth, such as:

1. Wireless control of communication between a mobile phone and a hands-free headset.
2. Wireless networking between PCs within the Bluetooth range.
3. Wireless connections between PC input and output devices, such as mouse, keyboard, and printers.
4. Bluetooth technology is used to replace the traditional communications in test equipments, medical equipments, bar code scanners, and traffic control devices.
5. Wireless control instead of infrared waves.
6. Transfer of files between devices with OBject EXchange (OBEX).
7. Transfer of contact details, calendar appointments, and reminders between devices with OBEX.
8. One of the most interesting applications of Bluetooth is establishing a WPAN within an office workplace.

## 1.6 Objective of the Book

This book aims at improving the performance of Bluetooth systems and the use of various error control codes. The different Bluetooth format is presented. The book studies the performance of Bluetooth systems over different communications channel such as AWGN and fading channels.

First, an analysis of the error probability in the standard Bluetooth frame format is presented. In addition, a study of the error control codes that are already used in the different parts of Bluetooth frame, the access code (AC), the header HD, and the payload, is performed. We focus in our study on the channel coding scheme of the payload in case of coded payload and uncoded payloads. Analyzing the Bluetooth performance will consider the additive white Gaussian noise (AWGN) and fading channel cases.

The book presents a modification for Bluetooth frame format to improve its performance over both AWGN and fading channels. Analysis of the effect of different block codes on the performance of Bluetooth system is presented. The book proposes the usage of other block codes such as the Hamming and the BCH codes with different lengths for the Bluetooth frame. It presents an analytical expression for Bluetooth Packet Error Probability (PEP) over a perfect interleaved fading channel in case of the standard Bluetooth frame and proposed frames.

Finally, the book proposes the usage of convolutional codes. for error correction in Bluetooth Convolutional codes the most widely used in communication systems. Also, modified Bluetooth packet formate is presented with the complete Matlab codes for the different simulation scenarios are introduced.

## 1.7 Outline of the Book

The main objective of the book is to improve simplify the performance of WPANs networks studying through the proposed Matlab codes and different scenarios are presented. Bluetooth system is chosen as a WPANs network for modelling and simulation. The book presents different proposed modifications of Bluetooth frame format through different channel coding schemes.

Chapter 2 presents a Bluetooth system description and the layers of Bluetooth system. The Bluetooth protocols, which serve this technology and network topologies, are presented. This chapter discusses the different frame contents and the physical links which are supported by this technology. Bluetooth packet types are also described in this chapter.

The role of channel coding in the Bluetooth frame, the types of channel codes, which are used in frame encoding, and brief descriptions of the block code are presented in Chap. 3. Chapter 3 also presents the different Bluetooth packet types with the error control codes, which are used, in each one of them.

An analysis of the Bluetooth system performance with standard error control codes over AWGN channels and fading channels is presented in Chap. 4. This analysis process is introduced for coded and uncoded Bluetooth packets. Chapter 4 also presents a modification of the Bluetooth frame format to improve its performance over both AWGN and fading channels by changing the channel codes in the a payload field by other block codes. This chapter investigates the effects of the proposed block codes on performance of the Bluetooth system. Also, it gives an analytical study of proposed format.

A brief description of convolutional codes is presented in Chap. 5. This chapter presents a modification for the Bluetooth frame format by using convolutional code type instead of block codes. An investigation of the effects of this modification on the performance of Bluetooth system over different channels is presented in this chapter.

Chapter 6 presents pseudo coding technique with Maltlab code simulation. The presented scenario proposes the use of chaotic maps in the Bluetooth communication systems for data encryption prior to transmission as a pseudo coding example. The purpose of this step is to reduce the effects of fading on the transmitted signals. Also, this chapter studies the effect of block coding on the performance of uncoded packet.

## References


1. IEEE 802.15 Working Group for WPANs. http://ieee802.org/15/
2. Lathi BP Modern digital and analog communication systems, 2nd ed. Holt, Rinehart and Winston
3. Mettala R (1999) Bluetooth protocol architecture: version 1.0, Bluetooth White Paper, Document # 1.C.120/1.0, 25 Aug 1999
4. Bluetooth Special Interest Group (2001) "Specifications of the Bluetooth System", Version 1.1, Jan 2001
5. Mikhalenko PV (2005) Developing wireless bluetooth applications in J2ME. JDJ 10(1). Retrieved on 1 July 2006

# Chapter 2
# WPANs Technologies Beginning

## 2.1 Introduction

Bluetooth is the world's new short-range RF transmission standard for small power consumption, low cost, short-range radio links between portable or desktop devices as shown in Fig. 2.1. This figure shows an example of a piconet form in the Bluetooth system. The Bluetooth technology aims to eliminate interconnection cables.

## 2.2 Bluetooth Characteristics

Bluetooth system provides a way to connect and exchange information between devices such as mobile phones, laptops, PCs, printers, and other digital devices over a secure globally unlicensed short-range radio frequency band. The Bluetooth specifications are developed and licensed by the Bluetooth Special Interest Group. Bluetooth systems attempt to provide significant advantages over other data transfer technologies, such as infrared (IrDA) and HomeRF (Radio Frequency), which are vying for similar markets. Despite the comments of Bluetooth Special Interest Group (SIG) indicating that the technology is complementary to IrDA, it is clearly a competitor for PC-to-peripherals connection. IrDA is limited in use by a short connection distance of 1 m and by the line-of-sight requirement for communication. So these limitations prevent IrDA to be used on a large scale [1].

Due to the Bluetooth RF nature, it is not subjected to these limitations and it can be used through walls and nonmetal objects. Bluetooth is a low cost, short distance, and easy to use system. Bluetooth handles both data and voice transmissions. It is capable of supporting asynchronous connectionless links for data transfer and synchronous connection-oriented links for voice transmission. So the Bluetooth system supports both synchronous and asynchronous physical connections.



M. A. M. El-Bendary, *Wireless Personal Communications*,
Signals and Communication Technology,
https://doi.org/10.1007/978-981-10-7131-7_2

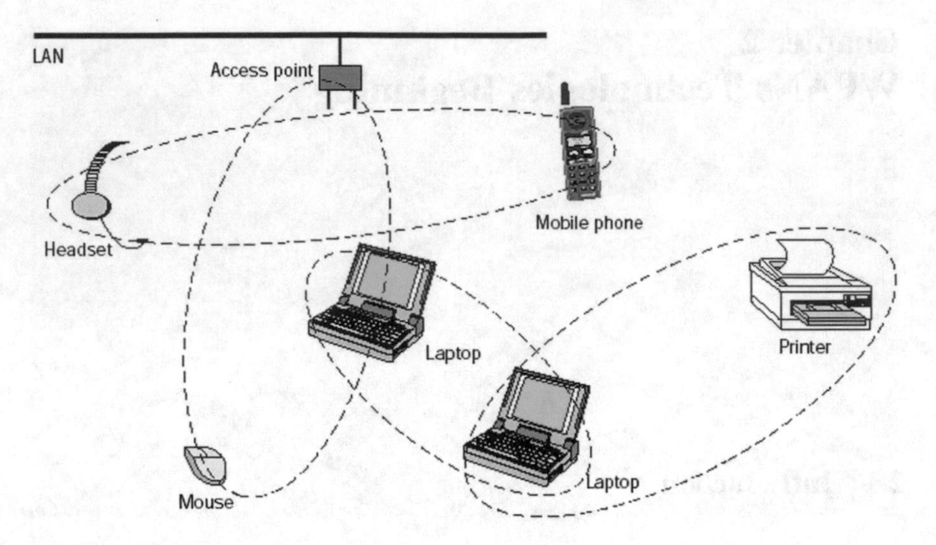


**Fig. 2.1** Scatternet with four piconets

## 2.3 Bluetooth System Architecture

The Bluetooth system stack is shown in Fig. 2.2. The first layer is the physical layer. It is the modem part. Inside this layer, the radio signal propagates. The second layer is the baseband layer. At this layer, the packets are formatted and the headers are created. In this layer also, check sum calculations are performed, the retransmission procedure is set, and the optionally encryption and decryptions steps are handled [2]. The entity that implements the baseband protocol and procedure is the link controller (LC). The Link Manager Protocol (LMP) is responsible for device setup links. Negotiations are used to set the power modes and duty cycles for the Bluetooth radio device. The logical link and Adaptation protocol (L2CAP) is responsible for reformatting the data packets into small units before it can be transmitted over the Bluetooth link. Also at the L2CAP layer, it is possible to set the Quality of Service (QoS) values for the link. The Host Controller Interface (HCI) is used for low-level hardware connection.

### *2.3.1 The Physical Layer*

In the physical layer, radio signals are processed by modems. The Bluetooth low-power radio operates in the unlicensed and globally available (ISM) band. The range of Bluetooth system extends from 2.4 to 2.4835 GHz. This band is globally available but the location and width of ISM band may vary from country to other country.

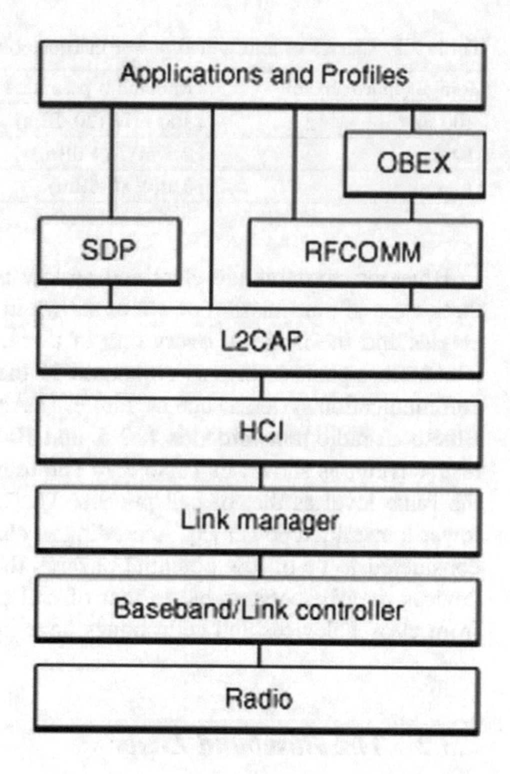


**Fig. 2.2** Bluetooth protocols stack

This band is shared by another communication system such as wireless LANs, cordless phones, and microwave ovens [2].

It is very difficult to predict the kind of interference in the ISM band. So interference cannot be neglected. Fortunately, the interference is limited to part of the ISM band only from 15 to 20 MHz which is occupied by microwave ovens signals [7]. To compact this interference, Bluetooth deploys a Frequency Hopping Spread Spectrum (FHSS) technology. In Bluetooth, there are 79 channels each with a 1 MHz bandwidth. The system has a hop rate of 1600 hops per second. That means that if one transmits on a bad channel, the next hop which is only 625 μs later, will be on another good channel. This faster hopping between frequencies improves can eliminate the interference at the expense of higher degree of complexity.

The hopping rate in Bluetooth is considered as a good trade-off between complexity and performance. The data rate in Bluetooth systems is 1 Mbps, but due to various kinds of protocols over head, it is reduced to be 723 Kbps. In Bluetooth systems, the signals are transmitting using Binary Gaussian Frequency Shift Keying (BGFSK) modulation. In recent Bluetooth versions, there are some new packet types which use other digital modulation.

**Table 2.1** Classes of transmitted power in Bluetooth system

| Range (approximate) | Maximum permitted power (mW/dBm) | Class |
|---|---|---|
| 100 m | 100 mW (20 dBm) | Class 1 |
| 10 m | 2.5 mW (4 dBm) | Class 2 |
| 1 m | 1 mW (0 dBm) | Class 3 |

Bluetooth systems are classified as low transmission power system. There are three class of transmission power as shown in Table 2.1. The table shows the power classes and the range of every one of them. In general, the transmitted power of Bluetooth signals is low as compared to the transmitted power of other wireless communication systems such as wireless LAN. The maximum power output from a Bluetooth radio transmitted is 1, 2.5, and 100 mW for class 3, class 2, and class 1, respectively, as shown in Table 2.1. The transmitted power of class 1 has roughly the same level as that of cell phones. The other two classes 2 and 3 have much lower transmitted power [3]. Accordingly, class 2 and class 3 Bluetooth devices are considered to be of low potential hazards than cell phones. But class 1 Bluetooth devices may be comparable to that of cell phones, where the transmitted powers from class 1 devices and cell phones have nearly the same value.

### 2.3.2 The Baseband Layer

There are two sub-layers in Bluetooth baseband layer: the lower baseband layer and upper baseband layer. In baseband layer, there are some processes performed such as packet formation, creation of packet headers, check sum calculations, retransmission, and encryption. The baseband layer is also responsible for the formation of the Bluetooth network (the piconet) between Bluetooth devices [4].

In the Bluetooth piconet, the role of Bluetooth devices is determined by baseband layer. A channel hopping sequence is used during traffic exchange within a piconet. This hopping sequence is unique for each piconet and it is determined by the Bluetooth device address (BD_ADDR) of the Master in piconet. The number of frequencies used for hopping is 79 frequencies. In the case of connection establishment, the number of hopping frequencies is 32 frequencies only.

The master of piconet starts its transmission in even-numbered hops only, but the slave starts the transmission in odd-numbered hops only. As shown in Fig. 2.3, Bluetooth packets can occupy more than one time slots. These packets are called multi-slot packets. The number of time slots is up to five time slots. In the case of multi-slot packets, the frequency of transmission is not changed with channel hopping sequence procedure, but it remains the same for the whole duration of the packet.

Bluetooth systems support two types of physical links. These links are Asynchronous Connectionless links (ACL) and Synchronous connection-Oriented SCO [7]. In Bluetooth systems, only a single ACL link can exist between a master

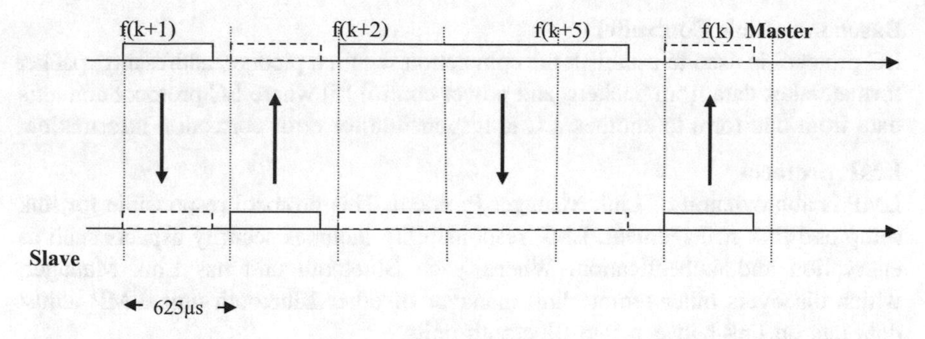


**Fig. 2.3** Channel hopping frequencies sequence

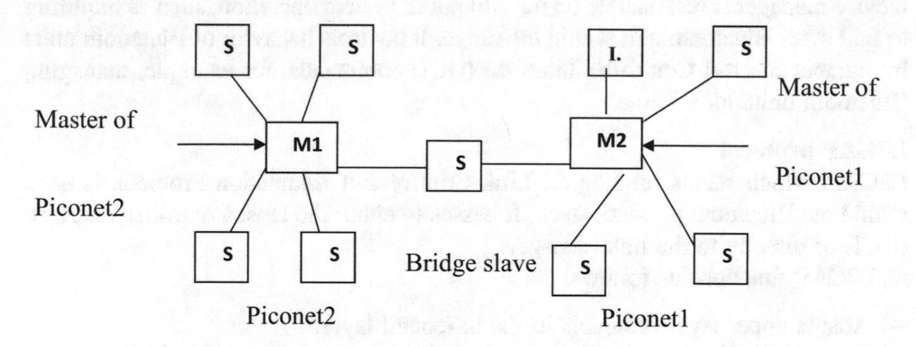


**Fig. 2.4** Bluetooth network structure (piconet)

and a slave in a piconet. Bluetooth supports up to three SCO links. A slave can support up to three SCO links from the same master or two SCO links from different masters. If the Bluetooth device is a member in more than one piconet, in this case Bluetooth device is called bridge slave as shown in Fig. 2.4.

Figure 2.4 shows two piconets forming a scatternet. The scatternet is formed by more than one piconet. The slave is a member in two piconets called a bridge slave where the data is exchanged between the two piconets through this member. Bluetooth system supports up to ten piconets within the Bluetooth distance range.

At least two piconets form scatternet as shown in the Figure. The number of members per piconet is up to seven slaves (S) plus one master (M).

### 2.3.3 *Bluetooth Protocols*

Bluetooth technology uses a variety protocols such as L2CAP, BNEP, RFCOMM, and SDAP. Each one of Bluetooth protocols has different role as follow [4].

**Baseband (Link Controller)**
LC protocol is used to establish the connection within a piconet, addressing, packet format, takes data from packets, and power control [5] where LC protocol converts data from one form to another. LC is responsible for error correction information.

**LMP protocol**
LMP is abbreviation of Link Manager Protocol. This protocol responsible for link setup and link management. LMP responsibility includes security aspects such as encryption and authentication. Where, each Bluetooth unit has Link Manager, which discovers other remote link manager of other Bluetooth unit. LMP adjust data rate on link between two Bluetooth units.

**The device manager (DM) protocol**
Device manager is responsible on the Bluetooth system operation, such as inquiring to find other Bluetooth unit within the range. It controls behavior of Bluetooth units by number of Host Controller Interface (HCI) commands, for example, managing Bluetooth unite local name.

**L2CAP protocol**
L2CAP, which stands for Logical Link Control and Adaptation Protocol, is used within the Bluetooth protocol stack. It passes to either the Host Controller Interface (HCI) or directly to the link manager.

L2CAP functions as follows:

- Adapts upper layer protocols to the baseband layer.
- Providing both connectionless (ACL) and connection-oriented SCO links.
- Multiplexing data between different higher layer protocols.
- Segmentation and reassembly of packets.
- Providing one-way transmission management to a group of other Bluetooth devices.
- Quality of service (QoS) management for higher layer protocols.

L2CAP used to communicate over the host ACL (Asynchronous Connectionless) link; its connection established after the ACL link has been set up (Fig. 2.5).

**HCI protocol**
HCI stands for Host Controller Interface. It is considered as standardized interface between the upper and lower layers in the Bluetooth communication system. It provides the capability of separating the radio H/W from higher layer protocol. Bluetooth module can be used for several different hosts and different applications by using HCI.

**BNEP protocol**
BNEP refers to Bluetooth Network Emulation Protocol, and it is used for delivering network packets on top of L2CAP. This protocol is used by the Personal Area Networking (PAN) profile.

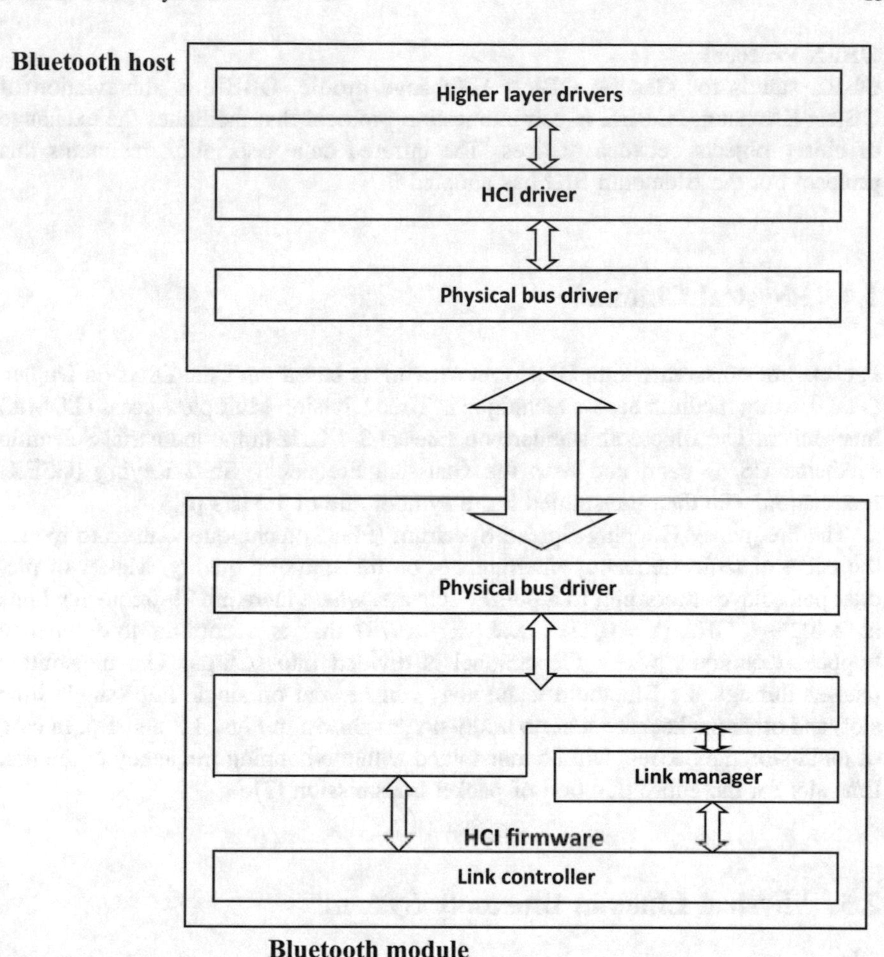


**Fig. 2.5** Bluetooth module

**RFCOMM protocol**
RFCOMM is abbreviation of Radio Frequency Communication. The Bluetooth protocol RFCOMM is a simple set of transport protocols, made on top the L2CAP protocol. This protocol provides emulated RS-serial ports, up to sixty simultaneous connections of Bluetooth devices at a time, where the Bluetooth serial port profile is based on this protocol.

**SDAP protocol**
SDAP refers to Service Discovery Application profile. It discovers the device information, services, and the characteristics if the services enable the establishment of the connection between the Bluetooth devices.

**OBEX Protocol**
OBEX stands for Generic OBject EXchange profile. OBEX is abbreviation of OBject EXchange. OBEX is communication protocol that facilitates the exchange of binary objects between devices. The infrared data association maintains this protocol but the Bluetooth SIG has adopted it.

## 2.4 Physical Channel

Packets transmission within Bluetooth network is based on Time Division Duplex (TDD), using medium access technique is Time Division Multiple Access (TDMA) mechanism. The Bluetooth standard operates at 2.4 GHz in the Industrial Scientific Medicine (ISM) band and uses the Gaussian Frequency Shift Keying (GSFK) modulation. The data transmitted has a symbol rate of 1 Ms/s [6].

The Frequency Hopping Spread Spectrum (FHSS) technique is used to reduce the effect of radio frequency interferences on transmission quality. Master of piconet polls slaves according to a polling scheme, where there are 79 frequency hops at 2.402 + k GHz (k = 0, 1, ……………, 78) that is according to frequency hopping technique, where the channel is divided into 625 μs. The transmitted packets through the Bluetooth technology can be sent on single hop (single time slot) and other packets are sent on multi-slot, as shown in Figs. 1.3 and 1.6. In case of multi-slot, the packets will be transmitted with the hopping frequency of the first time slot for the entire duration of packet transmission [7].

## 2.5 Physical Links in Bluetooth System

Bluetooth system supports two physical links: Asynchronous Connectionless (ACL) and Synchronous Connection-Oriented (SCO).

### *2.5.1 Asynchronous Connectionless (ACL) Link*

**ACL** link is used for data transmission with in Bluetooth system. ACL link provides a packet-switching communications between master and a slave, where it supports point-to-multipoint connection. ACL link is symmetric and asymmetric packet-switching connection. Within ACL link, an ARQ scheme is used for sure data integrity. By using the ACL packet, data is transmitted over one, three, or five time slots [8]. Data over ACL link may be protected by Forward Error Connection or with no FEC that is according to the packet type. With ACL link, a slave can support one ACL link with a master. Figure 2.6 shows a piconet of one master with

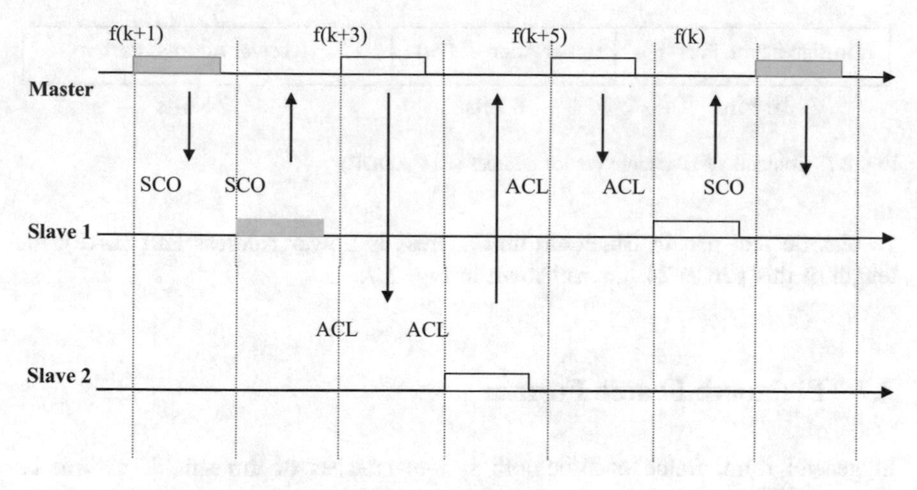


**Fig. 2.6** Physical link SCO and ACL within a piconet

one master and two slaves where slave 1 supports both SCO, which is a six time slots interval, and ACL links. Slave 2 supports ACL link only.

### 2.5.2 *Synchronous Connection-Oriented (SCO) Link*

Second physical link is SCO. SCO link is used for real-data transmission (voice) between Bluetooth units. SCO is a symmetric point-to-point connection between a master and slave. SCO link can be considered a circuit-switching connection, where the master sends SCO packets regular intervals. SCO packets occupy two, four, or six time slots. With SCO packets, the ARQ schemes are not used. A slave can support up to three SCO links from the same master or two SCO links from different masters.

## 2.6 Bluetooth Device Address

Every one of Bluetooth units has a unique physical address. Physical address of Bluetooth device (BD_ADDR) is 48 bits long, and it uniquely identifies any Bluetooth units. The Bluetooth device address is used to calculate access code (AC), where the AC is the main part in Bluetooth frame format, as shown in following section. Figure 2.7 shows the structure of Bluetooth device address. The address of Bluetooth units is comprises from two fields, first part is Upper Address Part (UAP), which is length an 8 bits and nonsignificant address part its length is

| Non-significant. Part | Upper Address Part | Lower Address Part |
|---|---|---|
| 16 bits | 8 bits | 24 bits |

**Fig. 2.7** Contents of Bluetooth device address (BD_ADDR)

16 bits. Second part in Bluetooth unit address is Lower Address Part (LAP); the length of this part is 24 bits as shown in Fig. 2.7.

## 2.7 Bluetooth Frame Format

In general form, frame of Bluetooth system consists of three fields as will be discussed. Figure 2.8 shows the frame format of Bluetooth.

### *2.7.1 Access Code (AC)*

**AC** is the first part in Bluetooth frame. AC length is 72 bits, as shown in Fig. 2.9, AC consists of 4 bits preamble, 64 bits synch word, and a 4 bit is trailer. Synch word is derived from LAP of Bluetooth unit address (24 bits). So the AC is unique every piconet where AC provides identification of all packets, which are exchanged within a piconet. Piconet is a collection of a number of Bluetooth unit up to 7 units with the same other Bluetooth unit is called a master of piconet. It provides synchronization between a master and slaves within a piconet. The access code (AC) is protected from channel errors [8].

| 72 Bits | 54Bits | 0-2745 Bits |
|---|---|---|
| Access code | Header | Payload |

**Fig. 2.8** Bluetooth frame format in general form

**Fig. 2.9** Access code (AC) format

| 4 Bits | 64 Bits | 4 Bits |
|---|---|---|
| Preamble | Synch word | Trailer |

Sync word is encoded within large Hamming distance $\geq 14$. AC uses error control code to resist channel errors, which can correct up to 6 errors. More details on this will be presented later. There are three different types of access code: Channel Access Code (CAC), Device Access Code (DAC), and Inquiry Access Code (IAC). Every one of these AC types carries out particular function as follows:

Channel Access Code (**CAC**): It is used to identify a piconet.

CAC is derived from master (M) Bluetooth unit address (BD_ADDR) of a piconet. It forms the preamble of all packets which are exchanged on the channel.

Device Access Code (**DAC**): It is used to identify a device and used for paging procedures. DAC is derived from paged Bluetooth unit address (BD_ADDR).

Inquiry Access Code (**IAC**): It is used for inquiry procedures.

IAC can be general (**GIAC**) or dedicated (DIAC), and there is one general (GIAC) and 63 dedicated IACs. GIAC is known to each Bluetooth device.

### 2.7.2 *Header of Bluetooth Frame (HD)*

Header part (HD) is second part in Bluetooth frame. It is 18 bits that follows the AC. The header HD is encoded with code rate 1/3 Forward Error Correction (FEC) and the output of encoder result is 54 bits header. Type of FEC and its features will be described later.

Figure 2.10 shows the HD contents with its different parts, each one of HD fields responsible for on determined function. The main function of the header of the Bluetooth packet is to determine an individual slave address in the piconet by Logical Transport Address (LT_ADDR) [11]. The different functions of HD fields are listed in the below of Fig. 2.10.

**LT_ADDR** is first field in HD, and it is three bits length; the maximum number of Bluetooth units within piconet is 7 Bluetooth units. So the field LT_ADDR length is restricted. It represents an active member address of Bluetooth units within a piconet.

**Type** it is 4 bits follows LT_ADDR. This field used to define the different Bluetooth packet types where there are 16 different packet types, by Type field the link type is determined ALC or SCO with its packet type and number of time slots which is occupied by packets.

**FLOW** it is single bit only used for flow control of Bluetooth packets over ACL link. If the receiving buffer is full, its bit value will be zero that means it stops transmission of packets. In case of its value one, then the packets transmission will be continued.

**ARQN** this field length single bit, it means acknowledgement indicates if the retransmission procedure is required or not, that is determined by the value of ARQN bit.

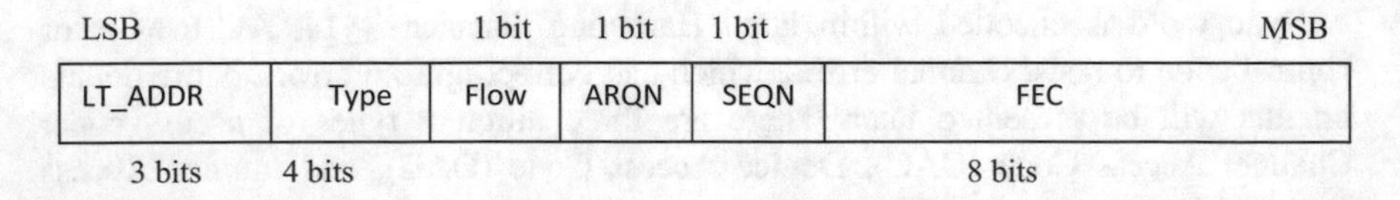


**Fig. 2.10** Header (HD) format

**SEQN** single bit is the length of this field, it provides sequential numbering to order the packet stream. That is required for filter out retransmissions at the Bluetooth device destination.

**HEC** it is the last field in HD; it means Header-Error-Check (HEC) checks header integrity.

### *2.7.3 Payloads*

The last section of Bluetooth frame is payload. Payload follows the HD part. The payload length is in range 0–2745 bits. That is payload length for basic Bluetooth packet types. The payload contents may be protected by Forward Error Correction (FEC) or not protected.

That is depending on the type of packets of Bluetooth. The payload part consists of three segments: first a payload header, a payload, and finally, 16-bit CRC for error detection. Number of time slots which are occupied by packets depend on the type of Bluetooth packet [9]. There are two types of packet according to time slots numbers. These are as follow:

Single time slot packet type.
Multi-time slots packets type.

## 2.8 Bluetooth Packet Types

There are many ways to classify the Bluetooth packets such as system packets or data packets. We can classify Bluetooth packet types according to its physical link type ACL or SCO link, the classification of packets may be according to number of time slots which is occupied by a packets. In Sect. (2.3.2), which describes the contents of packet header of Bluetooth HD, there is a second field that is called TYPE. The length of this field is 4 bits; this field allows for 16 different Bluetooth packet types for each of the link types. There are 4 packets for the Bluetooth system is called control packets; these packets are ID, NULL, POLL, and FHS, which are used for piconet setup [10].

**Table 2.2** ACL Bluetooth packet types

| Packet type | Slots and data bytes | Total bits |
|---|---|---|
| DM1 | 1 slot, 1-byte payload header, 17 bytes of data | 240 |
| DM3 | 3 slots, 2-byte payload header, 121 bytes of data | 1500 |
| DM5 | 5 slots, 2-byte payload header, 224 bytes of data | 2745 |
| DH1 | 1 slot, 1-byte payload header, 27 bytes of data | 240 |
| DH3 | 3 slots, 2-byte payload header, 183 bytes of data | 1496 |
| DH5 | 5 slots, 2-byte payload header, 339 bytes of data | 2744 |

Table 2.2 shows ACL packet types in Bluetooth system. All of packet types within Table 2.2 are payload packets. As shown in table, number of occupied time slots by packets and payload length also are different according to the type of packets.

DM1 and DH1 packets occupy one time slot.
DM3 and DH3 packets occupy three time slots.
DM5 and DH5 packets occupy five time slots.
DM data packets are protected by FEC schemes.
DH data packets are not protected.

**DM** and **DH** packets are used to carry asynchronous traffic; DM refers to data–medium. DM1 packet contains up to 18 data bytes. Data and CRC bits are encoded with rate 2/3 FEC. DH1 packets can carry up to 28 data bytes, and this packet is not used for FEC schemes. DH refers to data–high rate. DM3 and DM5 are considered as an extension of DM1. Similarly, DH3 and DH5 are extended DH1 packets.

High-quality voice (**HV**) packets are used for voice transmission, where voice packets are not retransmitted. HV1, HV2, and HV3 are used for voice transmission only but the last packet type is DV, which is used for transmission voice and data within two fields. Voice over Bluetooth network is transmitted using SCO link. Voice packets are not used for ARQ schemes, so it cannot be retransmitted. HV1 and HV2 packets are protected by FEC, but HV3 is not protected. HV2 packets occupy two time slots (Table 2.3).

HV2 packets occupy four time slots.
HV3 packets occupy six time slots.
DV voice/data fields.

**Table 2.3** SCO Bluetooth packet types

| Packet type | Time slots | Voice length |
|---|---|---|
| HV1 | 2 slots | 10 bytes, 1.25 ms voice |
| HV2 | 4 slots | 20 bytes, 2.5 ms voice |
| HV3 | 6 slots | 30 bytes, 3.75 ms voice |
| DV | 4 slots | 80 bits voice field and up to 150 bits data field |

## 2.9 Bluetooth Network Topologies

Bluetooth technology is based on master-slave concept where the master device controls data transmissions through connection setup and polling procedure. The master is defined as the device, which initiates the connection. A master Bluetooth device can communicate with up to seven Bluetooth units. There are two types of Bluetooth network topology: piconet and scatternet as shown below [9].

### *2.9.1 Piconet*

Bluetooth devices that are within the range can be connected together to set up a piconet. The number of Bluetooth units is connected together within piconet is up to seven slaves plus one master. A piconet is an ad hoc computer network, using Bluetooth technology protocols to allow one master device to interconnect with up to seven active devices that is Bluetooth devices collection called a piconet. Up to 255 further devices can be inactive, or parked, which the master device can bring into active status.

No more than eight Bluetooth units can be connected in single piconet. The master unit controls traffic in the network; every unit in a piconet network takes master identity and clock to track the hopping channel. Data within piconet can be transferred between the master and any other slave devices. In piconet, the devices can switch roles and the slave can become the master.

**Ad hoc network definition**
A wireless ad hoc network is computer network with wireless links between the network nodes, ad hoc means the connections are established when needed, ad hoc networks need simple configuration to establish a connection between its nodes. Nodes of ad hoc network are moving units. So ad hoc network should be self-configuring and self-healing.

### *2.9.2 Scatternet*

Bluetooth system allows connecting two or more piconets to form a scatternet. In scatternet construction, there is single or more than a Bluetooth devices acting as a bridge. A bridge device may be a master of a piconet and a slave member within other piconet. There is situation for bridge device, which can be a slave member within, two or more than piconets. Within scatternet, each piconet has unique identification, which is produced from address device (BD_ADDR) of master device of each piconet. That means each piconet chooses a different channel hopping sequence for packets transmission between its members.

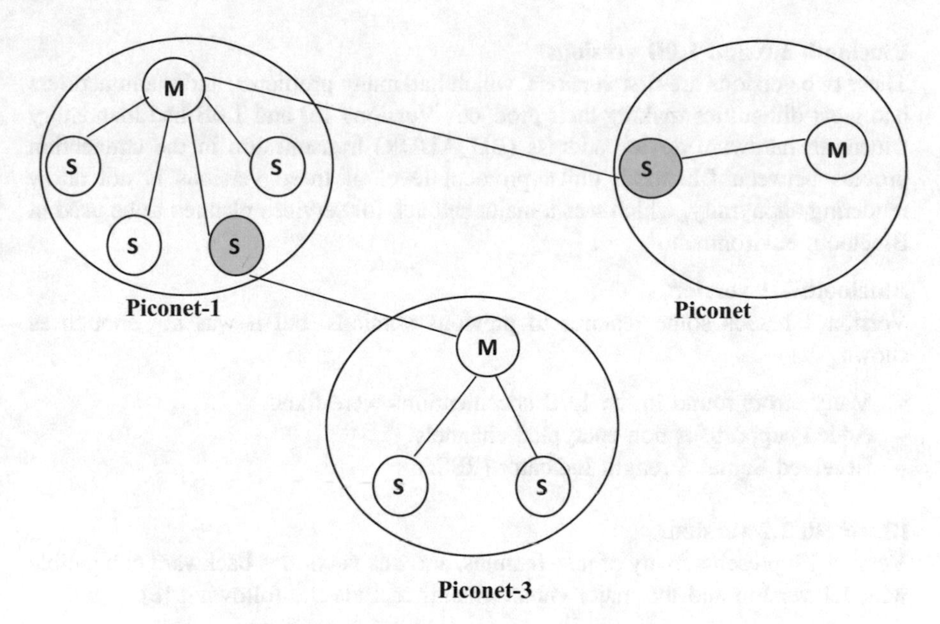


**Fig. 2.11** Scatternet formed from three piconets

Figure 2.11 shows a scatternet of Bluetooth system, which are conformed by three piconets. Each one of piconet has its mater; three piconets are connected together through its bridge slaves as shown in the figure. In other scatternet formation, a master of any piconet can be slave member in another one.

## 2.10 Bluetooth Versions Features and Specifications

Since the Bluetooth specification was developed in 1994, by Sven Mattisson and Haartsen. The specification of Bluetooth is based on frequency hopping spectrum technology (FHS). Bluetooth system operates in ISM band at 2.4–2.24835 GHz, and this frequency range suffers from many interference sources, which are coming from other wireless technologies. To avoid interfering with these technologies that use the 2.45 GHz, the Bluetooth system uses FHS, where the Bluetooth frequency band is divided to 79 channels each one of channels has 1 MHz bandwidth and changes channels up to 1600 hops per second.

There are many versions of Bluetooth that are aimed to improve data rate transmission and improve Bluetooth system features in general. Implementations with versions 1.1 and 1.2 data rate is 723.1 Kbps, data rate is increased with version 2.0 implementation feature Bluetooth enhanced data rate (EDR), where data rate reaches to 2.1 Mbps. As follows, the different Bluetooth versions are presented [11].

**Bluetooth 1.0 and 1.0B versions**

These two versions are first versions, which had many problems, and manufacturers had great difficulties making their products. Versions 1.0 and 1.0B had mandatory Bluetooth hardware device address (BD_ADDR) transmission in the connection process between Bluetooth units, protocol level of these versions is not allow rendering anonymity, which was a major setback for services planned to be used in Bluetooth environment.

**Bluetooth 1.1 version**

Version 1.1 adds some features to previous versions, but it was not enough as shown

- Many errors found in the 1.0B specifications were fixed.
- Added support for non-encrypted channels.
- Received Signal Strength Indicator (RSSI).

**Bluetooth 1.2 version**

Version 1.2 presents many of new features, and this version is backward compatible with 1.1 version and the major enhancements include the following [8]:

- Faster connection and discovery.
    - Adaptive frequency hopping spread spectrum (AFHSS), which improves resistance to radio frequency interference by avoiding the use of crowded frequencies in the hopping sequence.

  Higher transmission speeds in practice, up to 721 Kbps, as in 1.1 versions:

- Extended synchronous connections (eSCO), which improve voice quality of audio links by allowing retransmissions of corrupted packets.
- Host Controller Interface (HCI) supports for three-wire UART.
- HCI accesses to timing information for Bluetooth applications.

**Bluetooth 2.0 version +EDR (Enhanced Data Rate)**

Version 2.0 specified on 2004, which is backward compatible with version 1.1. The main enhancement more than version 1.1 is the introduction of an EDR of 3.0 Mbps [11]. EDR has many effects, as the following:

- Three times faster transmission speed up to 10 times in certain cases (up to 2.1 Mbps). The practical data rate is 2.1 Mbps and the basic signaling rate is about 3 Mbps [12].
- Lower power consumption through a reduced duty cycle.
- Simplification of multi-link scenarios due to more available bandwidth.

**Bluetooth 2.1 version +EDR (Enhanced Data Rate)**

Bluetooth version 2.1 +EDR specification is fully backward compatible with Bluetooth version 1.1, and it was adopted by the Bluetooth SIG on 2007. This specification includes the following:

- Extended inquiry response: It provides more information during the inquiry procedure to allow better filtering of devices before connection. This information includes the name of the device, a list of services the device supports, as well as other information like the time of day, and pairing information.
- Sniff sub-rating: It reduces the power consumption when devices are in the sniff low-power mode, especially on links with asymmetric data flows. Human Interface Device (HID) is expected to benefit the most, such as mouse and keyboard devices increasing the battery life by factor 3–10 [13].
- Secure Simple Pairing: It improves the pairing experience for Bluetooth devices, while increasing the use and strength of security. It is expected that this feature will significantly increase the use of Bluetooth.

**Future of Bluetooth**
In future, it expected the field of using Bluetooth will increase. The transmission speeds of data will increase. Bluetooth version 3.0 will be the next version of Bluetooth after version 2.1.

## 2.11 Bluetooth Security

Security has played a major role in the invention of Bluetooth. The Bluetooth SIG has put much effort into making Bluetooth a secure technology and has security experts who provide critical security information. In general, Bluetooth security is divided into three modes:

(1) Non-secure,
(2) Service-level enforced security, and
(3) Link-level enforced security.

In non-secure, a Bluetooth device does not initiate any security measures. In service-level enforced security mode, two Bluetooth devices can establish a non-secure Asynchronous Connectionless (ACL) link. Security procedures, namely authentication, authorization, and optional encryption, are initiated when a Logical Link Control and Adaptation Protocol (L2CAP) Connection-Oriented or Connectionless channel request is made [14, 15].

Bluetooth's security procedures include authorization, authentication, and optional encryption. Authentication involves proving the identity of a computer or computer user, or in Bluetooth's case, proving the identity of one piconet member to another. Authorization is the process of granting or denying access to a network resource. Encryption is the translation of data into secret code. It is used between Bluetooth devices so that eavesdroppers cannot read its contents. However, even with all of these defense mechanisms in place, Bluetooth has shown to have some security risks [16].

## References


1. Bluetooth specifications v1.1. http://www.bluetooth.com
2. Specification of bluetooth system, volume 2, version 1.1, February 22, 2001
3. Bluetooth specifications v1.2. http://www.bluetooth.com
4. Dasgupta K. Bluetooth protocol, and security architecture review. Online report, http://www.cs.utk.edu/~dasgupta/bluetooth/
5. Muller T (1999) Bluetooth security architecture: version 1.0. Bluetooth white paper, Document # 1.C.116/1.0, July 15, 1999
6. Bluetooth specifications core v2.0. http://www.bluetooth.com
7. Haartsen JC (2000) The bluetooth radio system. IEEE Pers Commun 7(1):28–36
8. IEEE 802.11, the working group setting the standards for wireless lans. http://grouper.ieee.org/groups/802/11
9. IEEE 802.15 wpan task group 1 (tg1). http://www.ieee802.org/15/pub/TG1.html
10. Candolin C. Security issues for wearable computing and bluetooth technology. Online report, http://www.cs.hut.fi/Opinnot/Tik-86.174/btwearable.pdf
11. Vainio JT (2000) Bluetooth security. Online report, May 25, 2000 http://www.niksula.cs.hut.fi/~jiitv/bluesec.html
12. Mohamed MAM, El-Azm AA, El-Fishawy NA, El-Tokhy MAR, Abd El-Samie FEA (2008) Optimization of bluetooth frame format for efficient performance. Prog Electromagn Res M, Vol. 1
13. Khalil AA, Elnaby MMA, Saad EM, Al-Nahari AY, Al-Zubi N, El-Bendary MAM, Abd El-Samie FEA (2014) Efficient speaker identification from speech transmitted over Bluetooth networks. Int J Speech Technol, Springer
14. El-Bendary MAM, Abou-El-Azm AE, El-Fishawy NA, Shawki F, El-Tokhy M, Abd El-Samie FEA, Kazemian HB (2013) Image transmission over mobile Bluetooth networks with enhanced data rate packets and chaotic interleaving. Wireless networks, Springer, Vol. 13
15. El-Bendary MAM (2017) FEC merged with double security approach based on encrypted image steganography for different purpose in the presence of noise and different attacks. Multimedia Tools and Appl, Springer
16. El-Bendary MAM (2015) Developing security tools of WSN and WBAN networks applications, Springer

# Chapter 3
# Error Control Schemes

## 3.1 Introduction

Error control codes add redundancy bits (called parity check bits) to the transmitted data (data word) to form codeword. With error control techniques utilizing, the receiver can detect only or detect and correct the errors. Error control codes are used in communications systems to protect the transmitted information from noise and interference. It also, reduces the number of errors in the received information. The channel coding technique inserts number of bits to the original data, these bits called check bits or redundant bits. These additional bits will allow only a capability of detection and correction of number of bit errors in received data. The number of detection and correction errors depends on the type of error control codes ant its parameters. The redundant bits lead to reduction of data rate transmission; this is the cost of using channel coding to protect the transmitted information [1].

## 3.2 Channel Codes Types

There are two types of channel codes

1. Block codes.
2. Convolutional codes.

There are many differences between block codes and convolutional codes. These two channel codes are used for detection and correction of channel bit errors. This chapter presents important terms of block codes in general, where block codes accept a block of $k$ information bits as input encoder and produce a block of $n$ information bits as output of encoder. There are many types of block codes which are commonly



M. A. M. El-Bendary, *Wireless Personal Communications*,
Signals and Communication Technology,
https://doi.org/10.1007/978-981-10-7131-7_3

used; these block codes are Hamming codes, Golay codes, and BCH codes. The error control schemes which are utilized in Bluetooth packet protection are block codes. The following block channel coding, BCH code, repetition code, and Hamming code are used for encoding the different section of Bluetooth packet [2].

In this chapter, convolutional code is not discussed here; it will be presented in Chap. 5. As shown in Fig. 3.1, there are important terms, which determine the rate of error control code and the number of redundant bits with the efficiency of error control codes [3].

*Important Terms' Definitions:*

Codeword (*n*): It is the output of encoder (*n* bits block), and it is known as codeword. It consists of the data word bits plus redundant bits.

$$n = k + m \tag{3.1}$$

Data word (*k*): It is the input of encoder, *k* bits block, and it is known as data word. It is the difference between codeword *n* and redundant bits *m*.

$$k = n - m \tag{3.2}$$

Code rate (*R*): It is the ratio of the input of encoder (*k*), data word to the output of encoder (*n*) codeword.

$$R = k/n \tag{3.3}$$

Code efficiency: This term is equal to the code rate (*R*).
Code efficiency = *k*/*n*
Error control codes have two methods to achieve its function:

1. Error detection with retransmission.
2. Error detection and correction.

In case of first type, after receiving the error will be detected, and so the receiver will send Automatic Repeat reQuest (ARQ) to transmitter. This method causes a lot of time wasting. In case of second type, the errors will be corrected after detection, hence the receiver will not send ARQ. The number of bits which are added by encoder = (bits at the output of encoder)—(bits at input of encoder) *m*, is shown in Fig. 3.1. The redundant bits cause error detection or detection and correction of error control codes capabilities [4].

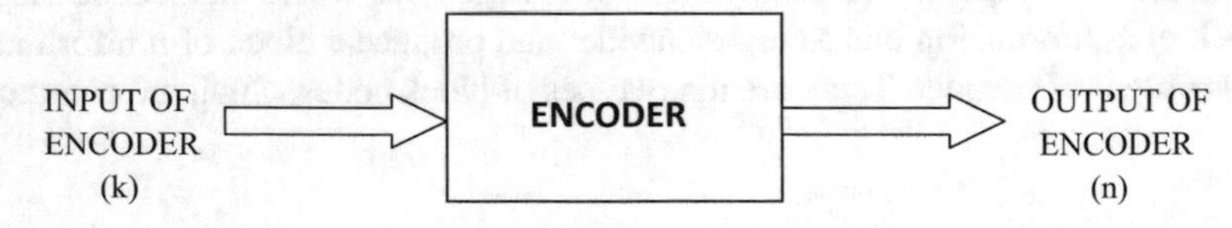


**Fig. 3.1** Encoder block, its input, and its output

## 3.3 Bluetooth Frame with Channel Coding

Bluetooth system operates in unlicensed 2.4 GHz industrial, scientific, and medical (ISM) band. This leads to the occurrence of much more errors. In Bluetooth system operation band, it suffers from noise and interference. So the channel coding is required in frame format of Bluetooth frame to detect and correct the number of bit errors. The frame format of Bluetooth is comprised of three parts, access code is first part, second part is header packet, and the last part is the payload. Each one of these parts uses different encoder as follows. Bluetooth system uses different types of error control block codes, and these block codes are different in its number of bit errors, which can be corrected in every block. Figure 3.2 shows the different format of data protocols' types [5].

**Access code (AC)**
AC is first part in Bluetooth frame. In this part, the type of channel code is Bose-Chaudhuri-Hocquenghem (BCH) with code rate (64, 30); this code can correct up to six errors. The BCH code is considered one of the most powerful classes of linear block codes. It provides Hamming distance $\geq (2t + 1)$. In this AC case, the codeword $n$ is 64-bit block and the data word, which is called sync word, is

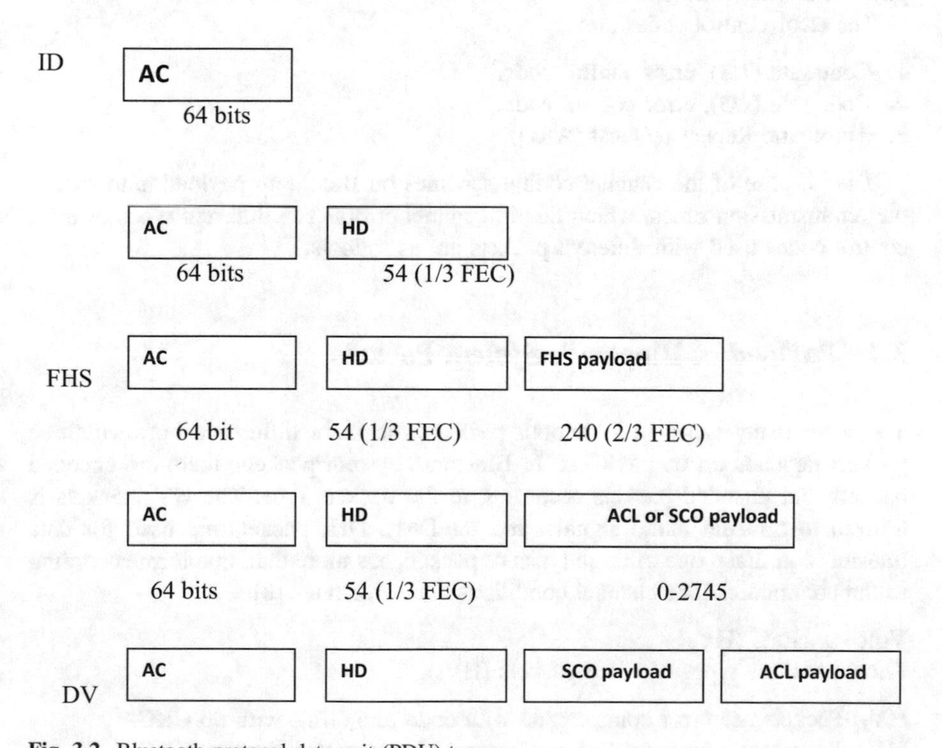


**Fig. 3.2** Bluetooth protocol data unit (PDU) types

30-bit block. The sync word of 30 bits consists of 24 bits; these 24 bits are lower part address (LPA) of Bluetooth device address (BD_ADDR) IEEE 802. The least six bits are added to improve the correlation properties at the receiver [6].

**Header of packet (HD)**
Header of Bluetooth frame is the second part in frame; HD length is 18 bits before encoding. HD is encoded by repetition code with code rate (3, 1). So the data word length is 1-bit block, and the codeword length is 3-bit block. Repetition code is simplest type of linear block code, where single bit is encoded into $n$ identical bits to produce ($n$, 1) block code. In case of HD, the number of $n$ bits is three bits. So the codewords in this code are an all-zeros codeword and an all-one codeword. The length of HD after encoding by repletion code (3, 1) will be 54 bits.

I. The payload

The payload part is the last part in Bluetooth frame. There are some packet types encoded as $DM_x$ packets and uncoded packets as $DH_x$ packets. In case of Bluetooth system payload, there are three types of error control codes, which are used in Bluetooth payload; the type of error control codes is used according to the type of payload as shown in [7].

The error control codes are:

1. Code rate (1/3), error control code.
2. Code rate (2/3), error control code.
3. Automatic Repeat reQuest (ARQ).

The purpose of the channel coding schemes on Bluetooth payload is to reduce the retransmission times, which due to channel errors. The different types of error control codes used with different packets are as follows.

## 3.4 Payload of Bluetooth System Packets

There are many types of Bluetooth packets; the main difference between these packets depends on its payloads. In Bluetooth system packets, there are encoded packets and uncoded packets according to the packets type. The $HV_x$ packets is utilized to transmit audio signals, and the $DM_x$, $DH_x$ packets are used for data transmission. Each one of these types of packets has more than one form according to the communications channel conditions and error rates [8].

**Voice packets ($HV_x$)**
There are three types of voice packets ($HV_x$).

**$HV_1$**: Packets use error control code with code rate (1/3), with no CRC.
**$HV_2$**: Packets use error control code with code rate (2/3), with no CRC.
**$HV_3$**: Packets uncoded packets (no FEC), with no CRC.

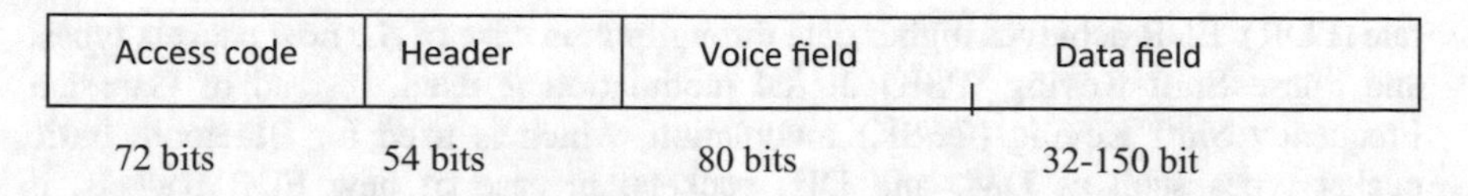


**Fig. 3.3** DV packet type of bluetooth system

**Data packets with medium rate ($DM_x$)**
There are three types of $DM_x$ packets. *M* symbol refers to medium rate. The three types of $DM_x$ packets are $DM_1$, $DM_3$, and $DM_5$, and each one of these packets uses the same error control code with rate (2/3), with 16-bit CRC.

**$DM_1$**: Packets use error control code with code rate (2/3), with 16-bit CRC.
**$DM_3$**: Packets use error control code with code rate (2/3), with 16-bit CRC.
**$DM_5$**: Packets use error control code with code rate (2/3), with 16-bit CRC.

**Data packets with high rate ($DH_x$)**
There are three types of $DH_x$ packets. *H* symbol refers to high rate.

**$DH_1$**: Packets uncoded packets but there is 16-bit CRC.
**$DH_3$**: Packets uncoded packets but there is 16-bit CRC.
**$DH_5$**: Packets uncoded packets but there is 16-bit CRC.

**There are other types of Bluetooth packets:**

**DV packets**
It is a packet combined from part data and voice as shown in Fig. 3.3.

In payload section, it is divided between voice part and data part. The length of voice part is 80 bits. Voice part is uncoded (no FEC) and no retransmission. The data part length is up to 150 bits with error control code (2/3) code rate and retransmission.

This type of packets is $DH_1$ packet, but its length is 30 bytes with no CRC. It means that this packet is uncoded packet.

In case of (real-time data) voice transmission, synchronous connection-oriented (SCO) links are used, like voice packets ($HV_x$). But in case of data transmission, asynchronous connectionless (ACL) links are used, such as data packets $DM_x$ and $DH_x$.

## 3.5 Bluetooth Enhanced Data Rate (EDR)

The previous types of packets, such as $DM_x$ packets and $DH_x$ packets, are called ACL packets. There are other data packet types, which are proposed in recent version of Bluetooth system; this new baseband packets are called enhanced data rate (EDR). EDR achieves higher data throughput, in case of the new packets types, and Phase Shift Keying (PSK) digital modulation is used, instead of Gaussian Frequency Shift Keying (GFSK) modulation, which is used for Bluetooth basic packet types such as $DM_x$ and $DH_x$ packets; in case of new EDR packets, it occupies 1, 3, or 5 time slots similar as basic Bluetooth packets, where each time slots are 625 microseconds in length, and the new EDR packets are considered to be ACL packets [9].

There are two types of new packets in EDR, each one of two EDR types depends on the type of PSK modulation, which is used in its case.

**First EDR packets**

It is called $2DH_1$, $2DH_3$, and $2DH_5$. In this case, the type of PSK modulation is Π/4-DQPSK modulation, which is employed for $2DH_x$ EDR packets. As shown in Table 3.1, the number of time slots per $DH_x$ packets in case of basic Bluetooth packets is not changed. The length of payload bits is increased due to the new modulation.

**Second EDR packets**

It is called $3DH_1$, $3DH_3$, and $3DH_5$. In this case, the type of PSK modulation is 8-DQPSK modulation, which is employed for $3DH_x$ EDR packets. Table 3.2 tabulates the number of time slots which are occupied by the $DH_x$ packets in as basic Bluetooth packets case. The length of payload bits is increased due to the new modulation [9].

So the new baseband Bluetooth packet types are considered as modified $DH_x$ packet types. With new EDR packets, the rest of Bluetooth frame remains the same, and the modulation of the header of packet is GFSK, such as basic Bluetooth packets ($DH_x$, $DM_x$) packets. In new specification, the symbol transmission rate (1 mega-symbol per second) remains unchanged but in the new EDR packets, the PSK modulation allows each symbol in the packet payload to carry more than single bit.

**Table 3.1** Bluetooth ACL packets (EDR) data types (Π/4-DPSK) payload modulation

| Packet type | Number of slots | Data length |
|---|---|---|
| $2DH_1$ | 1 slot | 54 bytes |
| $2DH_3$ | 3 slots | 367 bytes |
| $2DH_5$ | 5 slots | 679 bytes |

**Table 3.2** Bluetooth ACL packets (EDR) data types (8-DPSK) payload modulation

| Packet type | Number of slots | Data length |
|---|---|---|
| $3DH_1$ | 1 slot | 83 bytes |
| $3DH_3$ | 3 slots | 552 bytes |
| $3DH_5$ | 5 slots | 1021 bytes |

In this section the **ACL** packets is discussed. The basic Bluetooth packets are:

$DH_x$ packets.

$DM_x$ packets.

There is a main difference between the basic Bluetooth packets and the new Bluetooth packet types (EDR), which is there are error control codes (no FEC) in case of new packets (EDR), that is unlike the basic Bluetooth packets, $DM_x$ packets.

The error control repetition code (1, 3) scheme is used for encoding the header (HD) part in Bluetooth packet. It can correct single error and detect up to two errors. This type of error control code is used with $HV_1$ packets.

The error control code (2/3) rate is shortened Hamming code (15, 10), which is employed to protect from transmission channel error. This code can correct single error and detect up to two errors of every 10-bit block length. Shortened Hamming code (15, 10) is used in case of the following packets $HV_2$, $DM_1$, $DM_3$, and $DM_5$. In this channel code, for each DM mode packet, each block of 10 information bits (input of encoder, $k$) is encoded into 15-bit block (output of encoder, $n$). So the length of data word $k = 10$ and the length of codeword $n = 15$, and the code rate = $k/n$. The use of error control code in $DM_x$ packets will enhance the transmission performance, where it permits correcting single error every transmitted block [10].

In this book, WPANs simulation using Matlab program is presented. In the simulation and the presented Matlab experiments scenarios, ACL packets is considered. ACL refers to asynchronous connectionless. There are two types of ACL packets such as encoded packets $DM_x$ and uncoded packets $DH_x$. The symbol $x$ denotes the number of time slots between two frequency hops. One time slot equals 625 s where $x$ takes one value from three 1, 3, or 5. As shown previously, $DM_x$ packets always are encoded packets, but $DH_x$ packets are uncoded packets [11].

## References

1. IEEE 802.15 Working Group for WPANs, http://ieee802.org/15/
2. Lathi BP Modern digital and analog communication systems, 2nd edition, Holt, Rinehart and Winston
3. Muller T (1999) Bluetooth security architecture: version 1.0., Bluetooth White Paper, Document # 1.C.116/1.0, 15 July 1999
4. Bluetooth specifications core v2.0, http://www.bluetooth.com
5. Wicker SB (1995) Error control system for digital communication and storang upper saddle river. Prentice-Hall, NJ
6. Chen L-J, Sun T, Sanadidi MY, Gerla M (2004) Improving wireless link Throughput via interleaved fec. In: The ninth IEEE symposium on computers and communications
7. El-Bendary MAM (2013) Mobility effects combating through efficient low complexity technique. Ciit- Digital Image Processing
8. El-Bendary MAM, El-Tokhy M WSN ZigBee Based Performance. Ciit—J Wirel Commun

9. Galli S, Famolari D, Kodama T (2004) Bluetooth: channel coding considerations. In: IEEE vehicular technology conference, 17–19 May 2004
10. Mohamed MAM, El-Azm AA, El-Fishawy NA, El-Tokhy MAR, Abd El-Samie FEA (2008) Optimization of bluetooth frame format for efficient performance. Prog Electromagn Res M 1:101–110
11. El-Bendary MAM, El-Azm AA, El-Fishawy NA, El-Tokhy MAR, El-Samie FEA (2013) Image transmission over mobile bluetooth networks with enhanced data rate packets and chaotic interleaving. Wireless networks, Springer

# Chapter 4
# WPAN-Bluetooth Simulation Scenarios Using Block Codes

## 4.1 Introduction

This chapter aims to analyze the effects of Bluetooth existing error control codes on its performance over Rayleigh-block fading and AWGN channels. Also, this chapter presents an analytical Packet Error Probability (PEP) of Bluetooth frame on a perfectly interleaved Rayleigh fading channel. As seen in previous chapter, the channel codes are used in Bluetooth frame encoding where the shortened Hamming code (15, 10) is used for a payload of encoded packets $DM_x$. Types of channel codes which are used for access code (AC) and header (HD) of Bluetooth frames are BCH (64, 30) code and repetition (3, 1) code, respectively [1].

## 4.2 Analytical Studying of PEP

In this section, analytical analysis of the encoded WPAN-Bluetooth packets performance have been presented. In the PEP of the encoded packet, the standard error control schemes of the WPAN-Bluetooth network are considered.

First, the theoretical study of the PEP for interleaved Rayleigh fading channel needs bit error probability expression. In the proposed simulation, which will be discussed later, there are some simulation assumptions such as type of digital modulation—Binary Phase Shift Keying (BPSK) instead of GFSK which is used in Bluetooth frame modulation. In analyatical studying of WPAN-Bluetooth performance, the BPSK digital modulation is considered in the PEP analytical studying of WPAN-Bluetooth system.

The theoretical expression of the bit error probability $P_b$ of uncoded BPSK over Rayleigh-flat fading channels and coherent detection is given by [2, 3]:


M. A. M. El-Bendary, *Wireless Personal Communications*,
Signals and Communication Technology,
https://doi.org/10.1007/978-981-10-7131-7_4

$$p_b = \frac{1}{2}\left(1 - \sqrt{\frac{E_b/N_0}{1 + E_b/N_0}}\right) \cong \frac{1}{4.E_b/N_0} \tag{4.1}$$

$P_b$ is bit error probability (BEP).

$E_b/N_0$ signal-to-noise ratio (SNR) in case of uncoded data.

The approximation holds for high $E_b/N_0$ typically $E_b/N_0 > 13$ dB. Other assumption used in this section is that bit errors are uniformly distributed along the packet. Each part from Bluetooth frame is encoded with different error control codes. First, the equation will be in general case—let us consider the packet length $L$ with an error control code and its code rate $R$ [2]

$$R = \frac{k}{n} \tag{4.2}$$

where

$k$ data word length that is the input of encoder $k$-bit block.
$n$ codeword length that is encoder output of $n$-bit block.

With assumption that this error control code can correct up to $t$ number of errors, $t$ = No. of capable of correcting errors.

So the length of transmitted packet can be known by ($L$), $L = m \times n$, and $m$ is the number of codeword in this packet. The codeword error probability, in general, is a function of $t$ which is the number of errors which can be corrected by the code, $n$ which is the codeword length, and $m$ which is the number of codeword. To find the error probability over the whole frame, we use the following equation [3]:

$$P_{ecw} = \sum_{i=t+1}^{n} \binom{n}{i} P_b^i (1 - P_b)^{n-i} \tag{4.3}$$

$P_b$: BER prior to decoding and
$P_{ecw}$: Error probability of codeword.

Note that Eq. (4.3) is only for the case of independent errors and not for the case of burst errors. $E_b/N_0$ is substituted by $E_s/N_0$, where $E_s$ is code symbol energy. For binary signaling $E_s/N_0$ will be related to $E_b/N_0$ as follows:

$$\frac{E_s}{N_0} = \frac{k}{n}\frac{E_b}{N_0} \tag{4.4}$$

Since the energy available per codeword transmission is $kE_b$ joules, this is distributed among $n$ code symbols. So the probability that a packet is dropped $P_{\text{Badpckt}}$ is given by:

$$P_{\text{Badpckt}} = 1 - P_{\text{Goodpckt}} = 1 - (1 - p_{\text{ecw}})^m \tag{4.5}$$

where $P_{\text{Badpckt}}$ means that there is at least one error in the packet after decoding. $m$ is the number of codeword within a packet.

Bluetooth frames contain three parts: the AC, the HD, and the payload. To find relation between (PEP) packets error probability as a function in SNR, we must take into our consideration the difference of error control which is adapted in different packet fields such as access code (AC), header (HD), and payload (PL). The PEP is given by Eq. (4.6):

$$\text{PEP} = 1 - \left(1 - p^{\text{AC}}\right)\left(1 - p^{\text{HD}}\right)\left(1 - p^{\text{PL}x}\right) \tag{4.6}$$

In Eq. (4.6), $p^{\text{AC}}$ refers to AC codeword error probability, $P^{\text{HD}}$ is header error probability, and $P^{\text{PL}x}$ is payload part error probability.

There are three different coding schemes which belong to Bluetooth frame parts so the error probability is calculated separately.

The analytical deviation of $p^{\text{AC}}$, $P^{\text{HD}}$, and $P^{\text{PL}x}$ requires the knowledge of codeword error probability $P_{\text{ecw}}$, which is a function of the number of correctable errors $t$, the codeword length ($n$), and ($m$) number of codeword within frame.

**Access Code (AC)**

Codeword is $n = 64$, $t = 6$, and $m$ is the number of codeword equal to 1, where in case of AC it is single data word input of encoder so there is a single codeword. By using Eq. (4.3), we get

$$P^{\text{AC}} = \sum_{i=7}^{64} \binom{64}{i} P_b^i (1 - P_b)^{64-i} \tag{4.7}$$

**Header (HD)**

Codeword is $n = 3$, $m = 18$, $t = 1$, $k = 1$ which is data word, and the packet length $= m \cdot n = 54$ which is length of HD part after decoding process.

$$P_{\text{ecw}}^{\text{HD}} = \sum_{i=3}^{3} \binom{3}{i} P_b^i (1 - P_b)^{3-i} \tag{4.8}$$

Equation (4.8) gives error probability of single codeword of HD. The probability that the HD is discarded after decoding is given by Eq. (4.9):

$$P^{\text{HD}} = 1 - \left(1 - p_{\text{ecw}}^{\text{HD}}\right)^{18} \tag{4.9}$$

**Payload (PL)**

In this part, the payload section of Bluetooth encoded basic packets ($DM_x$) is discussed with its encoding process description. Basic Bluetooth packets have two types: $DM_x$ and uncoded $DH_x$ packets. The error control technique which is used for encoding the payload section of $DM_x$ packets is the shortened Hamming code (15, 10). It can correct single error per 10-bit block (data word length) and detect up to two errors in data word. There are three types of $DM_x$ as mentioned in previous chapter. These three types of $DM_x$ are different in its payload length. So the number of codeword of PL will be different as in the following equations [4].

In this case, we note that:

$n = 15$ bits,
$k = 10$ bits, and
$t = 1$.

So Eq. (4.3) in case of (15, 10) shortened Hamming code is given in Eq. (4.10) [17]:

$$P_{\text{ecw}}^{\text{PL}} = \sum_{i=2}^{15} \binom{15}{i} p_{\text{b}}^{i} (1 - p_{\text{b}})^{15-i} \tag{4.10}$$

From Eq. (4.10), we get the error of codeword probability for payloads. It is the same for all $DM_x$ packets, where its parameters do not depend on number of codeword $m$.

**$DM_1$ Packets**

$DM_1$ packet is one type of $DM_x$ packets. Length of its payload is $L = 160$ bits. So we note that $m = 16$ codeword. Error probability of whole $DM_1$ payload is given by Eq. (4.11):

$$P^{\text{PL1}} = 1 - \left(1 - P_{\text{ecw}}^{\text{PL}}\right)^{16} \tag{4.11}$$

**$DM_3$ Packets**

$DM_3$ packet is second type of $DM_x$ packets. Length of its payload is $L = 1000$ bits. So we note that $m = 100$ codeword. Error probability of whole $DM_3$ payload is given by Eq. (4.12):

$$P^{\text{PL3}} = 1 - \left(1 - P_{\text{ecw}}^{\text{PL}}\right)^{100} \tag{4.12}$$

**$DM_5$ Packets**

$DM_5$ packet is third type of $DM_x$ packets. Length of its payload is $L = 1830$ bits. So we note that $m = 183$ codeword. Error probability of whole $DM_5$ payload is given by Eq. (4.13).

$$P^{\text{PL5}} = 1 - \left(1 - P_{\text{ecw}}^{\text{PL}}\right)^{183} \tag{4.13}$$

From Eqs. (4.1, 4.3), we conclude that PEP is a function of SNR as all terms are functions of SNR.

The packet is accepted if AC, HD, and PL are fine. By substituting in Eq. (4.6), we get the PEP of Bluetooth packets as in Eq. (4.14) in general form for all $DM_x$ packets

$$\mathrm{PEP} = 1 - \left(1 - p^{\mathrm{AC}}\right)\left(1 - p^{\mathrm{HD}}\right)\left(1 - p^{\mathrm{PL}x}\right) \tag{4.14}$$

In case of $DM_1$ packets, PEP is given in Eq. (4.15):

$$\mathrm{PEP} = 1 - \left(1 - p^{\mathrm{AC}}\right)\left(1 - p^{\mathrm{HD}}\right)\left(1 - p^{\mathrm{PL1}}\right) \tag{4.15}$$

In case of $DM_3$ packets, PEP is given in Eq. (4.16).

$$\mathrm{PEP} = 1 - \left(1 - p^{\mathrm{AC}}\right)\left(1 - p^{\mathrm{HD}}\right)\left(1 - p^{\mathrm{PL3}}\right) \tag{4.16}$$

In case of $DM_5$ encoded packets, the PEP is given in Eq. (4.17):

$$\mathrm{PEP} = 1 - \left(1 - p^{\mathrm{AC}}\right)\left(1 - p^{\mathrm{HD}}\right)\left(1 - p^{\mathrm{PL5}}\right) \tag{4.17}$$

It is noted from previous equations that parts which belong to AC and HD are not changed with different types of $DM_x$. By using Eqs. (4.15, 4.16, and 4.17), we calculate the value of the theoretical PEP as functioned in SNR for $DM_1$, $DM_3$, and $DM_5$ packets. These probabilities of $DM_x$ packet error are plotted in Fig. 4.1 over perfect interleaved channel [5].

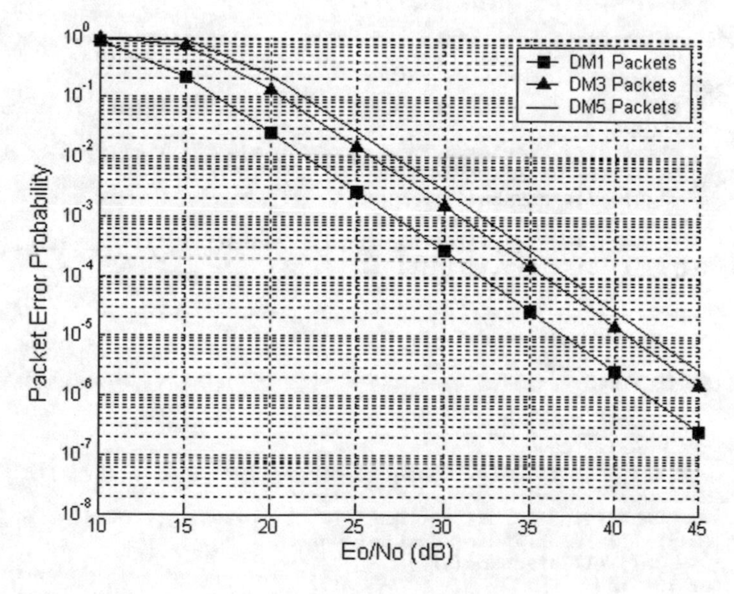


**Fig. 4.1** Analytical PEP of $DM_x$ packets (interleaved channel case)

#Matlab Codes of the DH*x* packets [6]#

```
% ========================IN THE NAME OF ALLAH =====================

% ====================ANALYTICAL ANALYSIS=====================

% ============RELATION BETWEEN THE SNR TO PACKET ERROR
PROBABILITY=======
clc
clear all
close all
k2=1;% input('the length of dataword header part = ');
k3=10;% input('the length of dataword payload part = ');
k1=30; %input('the length of dataword access code part = ');
  n1=64;%input('the length of codeword access code part = ');
  n2=3; %input('the length of codeword header part = ');
  n3=15; %input('the length of codeword payload part = ');

%snr=[10;15;20;25;30;35;40;45;50];

%%%%%%%%%%%%%%%%%%%%%%%%%%%%%%%%%%%%%%%%%%%%%%%%%%%%%%%%%%%%%%%%%%%%%%%%%
%%%
   snr=[10:5:45];
   %snr=[0:1:25];
y=length(snr);

for j=1:y
    snrabs(j)=10^(snr(j)/10);

      %####### For AWGN BEP
     %A(j)=sqrt(snrabs(j));
      % p(j)=0.5*erfc(A(j));
      %######for Fading channel
       %A(j)=1/snrabs(j);
       % B(j)=1+A(j)
       %p(j)=0.5*(1-(1/sqrt(B(j))));

%%%%%%%%%%%%%%%%%%%%%%%%%%%%%%%%%%%%%%%%%%%%%%%%%%%%%%%%%%%%%%%%%%%%%%%%%
%%%
    pew1(j)=0;
      % A1(j)=sqrt((k1/n1)*snrabs(j)); p1(j)=0.5*erfc(A1(j));% For
AWGN Channel
       %A1(j)=1/((k1/n1)*snrabs(j)); B1(j)=1+A1(j); p1(j)=0.5*(1-
(1/sqrt(B1(j))));  %For FAding Channel
       p1(j)=1/(4*snrabs(j));
 for i1=7:n1
         pew1(j)=pew1(j)+(factorial(n1)/(factorial(i1)*factorial(n1-
i1)))*((p1(j)^i1)*((1-p1(j))^(n1-i1)));

            end
             pacbad(j)=pew1(j);

%%%%%%%%%%%%%%%%%%%%%%%%%%%%%%%%%%%%%%%%%%%%%%%%%%%%%%%%%%%%%%%%%%%%%%%%%
%%%

           pew2(j)=0;
         %A2(j)=sqrt((k2/n2)*snrabs(j)); p2(j)=0.5*erfc(A2(j));% For
AWGN Channel
          %A2(j)=1/((k2/n2)*snrabs(j)); B2(j)=1+A2(j); p2(j)=0.5*(1-
(1/sqrt(B2(j))));  %For FAding Channel
          p2(j)=1/(4*snrabs(j));
     for i2=2:n2
```

```
pew2(j)=pew2(j)+(factorial(n2)/(factorial(i2)*factorial(n2-
i2)))*((p2(j)^i2)*((1-p2(j))^(n2-i2)));

            end
            pbad(j)=1-((1-pew2(j))^18);

%%%%%%%%%%%%%%%%%%%%%%%%%%%%%%%%%%%%%%%%%%%%%%%%%%%%%%%%%%%%%%%%%%%%%%%%%
%%%%%%%%%%%%%%%%%%%%%%%%%%%%%%%%%%%%%%%
            % pew31(j)=0;
            %A3(j)=sqrt(snrabs(j)); p3(j)=0.5*erfc(A3(j));% For AWGN
Channel


                 %A3(j)=1/snrabs(j);B3(j)=1+A3(j); p3(j)=0.5*(1-
     (1/sqrt(B3(j)))); %Fading channel
                 p3(j)=1/(4*snrabs(j));
        % for i3=2:n3

               %
     pew31(j)=pew31(j)+(factorial(n3)/(factorial(i3)*factorial(n3-
     i3)))*((p(j)^i3)*((1-p(j))^(n3-i3)));

          %end
            pbaddh1(j)=(1-p3(j))^240;
            pbaddh3(j)=(1-p3(j))^1500;
            pbaddh5(j)=(1-p3(j))^2745;
        pbaddh11(j)=1-((1-pacbad(j))*(1-pbad(j))*(pbaddh1(j)));
        pbaddh33(j)=1-((1-pacbad(j))*(1-pbad(j))*(pbaddh3(j)));
        pbaddh55(j)=1-((1-pacbad(j))*(1-pbad(j))*(pbaddh5(j)));
        snr;
        DH1(j)=pbaddh11(j);
        DH3(j)=pbaddh33(j);
        DH5(j)=pbaddh55(j);

        %############  Throughput  Calculationb  #############

        TH_per_dh1(j)=(240*(1-DH1(j)))/((1+1)*625);%
        TH_per_dh3(j)=(1500*(1-DH3(j)))/((3+1)*625);%
        TH_per_dh5(j)=(2745*(1-DH5(j)))/((5+1)*625);%

    end

    %disp(['the amount of ,pbaddm1=:',num2str(pbaddm1)]);
    %disp(['the amount of ,pbaddm3=:',num2str(pbaddm3)]);
    %disp(['the amount of pbad5 =:',num2str(pbaddm5)]);
    %semilogx;
    %semilogy(x);
    hold on;
    %grid on
    %snr=[10 15 20 25 30 35 40 45 50];
    semilogy(snr,DH1,'b-s');
    semilogy(snr,DH3,'k+:');
    semilogy(snr,DH5,'r+:');
    legend('DH1 Packets','DH3 Packets','DH5 Packets');
    title('PEP to SNR Classic DHx Packets over AWGN Channel');
    xlabel('Eo/No (dB)');
    ylabel('Packet Error Probability (PEP)');
    grid on
```

```
figure(60);

        semilogy(DH1,TH_per_dh1,'b-s');
        hold on
        semilogy(DH3,TH_per_dh3,'k+:');
        hold on
        semilogy(DH5,TH_per_dh5,'r+:');

       legend('THROUGHPUT DH1','THROUGHPUT DH3','THROUGHPUT DH5');
        title('THROUGHPUT to PER of Classic DHx Packets over AWGN
Channel');
        xlabel('Packet Error Probability (PEP)');
        ylabel('Throughput Mbps');
        grid on
        figure(70);
        semilogy(snr,TH_per_dh1,'b-s');
        hold on
        semilogy(snr,TH_per_dh3,'k+:');
        hold on
        semilogy(snr,TH_per_dh5,'r+:');
        hold on;
        legend('THROUGHPUT DH1','THROUGHPUT DH3','THROUGHPUT DH5');
        title('THROUGHPUT to SNR of Classic DHx Packets over AWGN
Channel');
         xlabel('SNR');
        ylabel('Throughput Mbps');
        grid on

        figure(80);

        semilogy(p3,TH_per_dh1,'b-s');
        hold on
        semilogy(p3,TH_per_dh3,'k+:');
        hold on
        semilogy(p3,TH_per_dh5,'r+:');

       legend('THROUGHPUT DH1','THROUGHPUT DH3','THROUGHPUT DH5');
        title('THROUGHPUT to BEP of Classic DHx Packets over AWGN
Channel');
        xlabel('Bit Error Probability (PEP)');
        ylabel('Throughput Mbps');
        grid on
```

# Matlab code of DMx packets#

```
% =====================ANALYTICAL ANALYSIS of Standard
BT==============

%  =====================RELATION BETWEEN THE SNR and
PEP================
clc
clear all
close all
```

```
k1=30; %input('the length of dataword AC (Access Code) part = ');
k2=1;% input('the length of dataword HD (Header) part = ');
k3=10;% input('the length of dataword payload part = ');
%#####################################################################
##
  n1=64;%input('the length of codeword AC (Access Code) part = ');
  n2=3; %input('the length of codeword HD (Header) part = ');
  n3=15; %input('the length of codeword PL (Payload) part = ');

%%%%%%%%%%%%%%%%%%%%%%%%%%%%%%%%%%%%%%%%%%%%%%%%%%%%%%%%%%%%%%%%%%%%%%
%%%%%%%%%%%%%%%%%%%%%%%%%%%%%%%%%%%%%
  snr=[10:5:50];
y=length(snr);

for j=1:y
    snrabs(j)=10^(snr(j)/10);

       p(j)=1/(4*snrabs(j));

%%%%%%%%%%%%%%%%%%%%%%%%%%%%%%%%%%%%%%%%%%%%%%%%%%%%%%%%%%%%%%%%%%%%%%
%%%%%%%%%%%%%%%%%%%%%%%%%%%%%%%%%%%
    pew1(j)=0;
    p(j)=1/(4*(k1/n1)*snrabs(j));
 for i1=7:n1
        pew1(j)=pew1(j)+(factorial(n1)/(factorial(i1)*factorial(n1-
i1)))*((p(j)^i1)*((1-p(j))^(n1-i1)));

            end
             pacbad(j)=pew1(j);


%%%%%%%%%%%%%%%%%%%%%%%%%%%%%%%%%%%%%%%%%%%%%%%%%%%%%%%%%%%%%%%%%%%%%%
%%%%%%%%%%%%%%%%%%%%%%%%%%%%%%%%%%%%%

             pew2(j)=0;
             p(j)=1/(4*(k2/n2)*snrabs(j));

       for i2=2:n2

pew2(j)=pew2(j)+(factorial(n2)/(factorial(i2)*factorial(n2-
i2)))*((p(j)^i2)*((1-p(j))^(n2-i2)));

             end
             pbad(j)=1-((1-pew2(j))^18);

%%%%%%%%%%%%%%%%%%%%%%%%%%%%%%%%%%%%%%%%%%%%%%%%%%%%%%%%%%%%%%%%%%%%%%
%%%%%%%%%%%%%%%%%%%%%%%%%%%%%%%%%%%%%%
             pew31(j)=0;
             p(j)=1/(4*(k3/n3)*snrabs(j));
     for i3=2:n3

pew31(j)=pew31(j)+(factorial(n3)/(factorial(i3)*factorial(n3-
i3)))*((p(j)^i3)*((1-p(j))^(n3-i3)));

     end
```

```
        pbaddm1(j)=1-((1-pew31(j))^16);
        pbaddm3(j)=1-((1-pew31(j))^100);
        pbaddm5(j)=1-((1-pew31(j))^183);
    pbaddm11(j)=1-((1-pacbad(j))*(1-pbad(j))*(1-pbaddm1(j)));
    pbaddm33(j)=1-((1-pacbad(j))*(1-pbad(j))*(1-pbaddm3(j)));
    pbaddm55(j)=1-((1-pacbad(j))*(1-pbad(j))*(1-pbaddm5(j)));
    snr;
    DM1(j)=pbaddm11(j);
    DM3(j)=pbaddm33(j);
    DM5(j)=pbaddm55(j);

end
%disp(['the amount of ,pbaddm1=:',num2str(pbaddm1)]);
%disp(['the amount of ,pbaddm3=:',num2str(pbaddm3)]);
%disp(['the amount of pbad5 =:',num2str(pbaddm5)]);
%semilogx;
%semilogy(x);

hold on;
%grid on
%snr=[10 15 20 25 30 35 40 45 50];
semilogy(snr,DM1,'b-s');
semilogy(snr,DM3,'k+:');
semilogy(snr,DM5,'r+:');
legend('DM1 Packets','DM3 Packets','DM5 Packets');
title('Eb/No TO PEP RELATIONS');
xlabel('Eo/No (dB)');
ylabel('(PEP)Packet Error Probability');
grid on
```

## 4.3 Proposed Frame Format and Analytical Studying

Type of error control codes, which is used within Bluetooth frames, is block codes. There are several different types of block codes. The previous section presented an analytical study for PEP that is for the existing Bluetooth system error control codes. This section presents an analytical study for PEP that is for Bluetooth frames but with other block codes over perfectly interleaved Rayleigh fading channel. We will use other block codes for payload of $DM_x$-encoded packets. The channel codes of AC and HD portions will be the same as Bluetooth frame standard. This analysis will show the effect of other block codes for Payload portion on Bluetooth system performance that is in analytically study. The proposed block codes in the following simulation scenarios are presented as follows [6]:

1. Hamming code (7, 4) and
2. BCH code (15, 7).

### 4.3.1 Hamming Code (7, 4)

In this section, different coding scheme is proposed to encode the payload field in Bluetooth frame. First, the Hamming code (7, 4) is proposed. This is accomplished by dividing the payload into four-bit segments, which are then encoded to 7-bit codeword with Hamming code (7, 4). This coding structure will be used for perfectly interleaved Rayleigh fading channel. In this case, the length of data word $k$ and codeword $n$ is changed. There is difference in $m$ which is number of codeword, as we will see in the following analysis [7].

To find the relation between PEP and the SNR, we must take into consideration the variants of error control codes which are proposed in this case for payload of $DM_x$ frames, as previously described in Sect. 4.2.

This PEP for Bluetooth frame is given in Eq. (4.6):

$$\text{PEP} = 1 - \left(1 - p^{\text{AC}}\right)\left(1 - p^{\text{HD}}\right)\left(1 - p^{\text{PL}x}\right)$$

where $P^{\text{AC}}$ and $P^{\text{HD}}$ are the same as mentioned previously, which are given in Eqs. (7, 9) without any changes where its error control codes are not changed. For the payload portion, $P^{\text{PL}x}$ is the payload error probability which will be changed as appeared in the following description.

The codeword error probability, in general, is a function of $t$ which is the number of errors which can be corrected by the code, $n$ which is the codeword length, and $m$ which is the number of codeword. To find the error probability over the whole frame, in case of using Hamming code (7, 4) for a payload of $DM_x$ frames, it is noted that the previous equations would be modified for the following variable:

$n = 7$ bits,
$k = 4$ bits, and
$t = 1$.

$P_{\text{Hecw}}$ is the error probability of payload codeword in case of Hamming code (7, 4), which is given by Eq. (4.18):

$$P_{\text{Hecw}}^{\text{PL}} = \sum_{i=2}^{7} \binom{7}{i} P_{\text{b}}^{i} (1 - P_{\text{b}})^{7-i} \qquad (4.18)$$

$P_{\text{BadPckt}}$ is the probability of packet error after decoding which is given in Eq. (4.5); it depends on $m$ which is codeword number.

For the Hamming code (7, 4), the length of the payload will be reduced to keep the Bluetooth frame with the same standard length and this modification makes the following:

**$DM_1$** payload = 136 bits → $m$ = 34.
**$DM_3$** payload = 856 bits → $m$ = 214.
**$DM_5$** payload = 1568 bits → $m$ = 392.

We note that the number of codewords has been changed as compared to shortened Hamming code (15, 10) and the length of payload decreased by 15%, and $t = 1$.

For Hamming code (7, 4), using Eq. (4.18) we get the codeword error probability.

From Eq. (4.18), we get the probability of error in $DM_1$, $DM_3$, and $DM_5$ packets as follows:

**For $DM_1$ Packets**

$DM_1$ packet is one type of $DM_x$ packets. Length of its payload is $\mathbf{L}$ = 136 bits. So we note that $m$ = 34 codeword. Error probability of encoded payload of $DM_1$ is given by Eq. (4.19):

$$P^{\mathrm{PLH1}} = 1 - \left(1 - p_{\mathrm{Hecw}}^{\mathrm{PL}}\right)^{34} \tag{4.19}$$

**$DM_3$ Packets**

$DM_3$ packet is second type of $DM_x$ packets. Length of its payload is $\mathbf{L}$ = 856 bits. So we note that $m$ = 214 codeword. Error probability of whole $DM_3$ payload is given by Eq. (4.20):

$$P^{\mathrm{PLH3}} = 1 - \left(1 - p_{\mathrm{Hecw}}^{\mathrm{PL}}\right)^{214} \tag{4.20}$$

**$DM_5$ Packets**

$DM_5$ packet is third type of $DM_x$ packets. Length of its payload is $\mathbf{L}$ = 1568 bits. So we note that $m$ = 392 codeword. Error probability of whole $DM_5$ payload is given by Eq. (4.21):

$$P^{\mathrm{PLH5}} = 1 - \left(1 - p_{\mathrm{Hecw}}^{\mathrm{PL}}\right)^{392} \tag{4.21}$$

AC and HD error probabilities $P^{\mathrm{AC}}$ and $P^{\mathrm{HD}}$ are given by Eqs. (4.7) and (4.9), respectively. We can conclude $\mathrm{PEP}^H$ of $DM_x$ packets by substituting in Eq. (4.6), where $\mathrm{PEP}^H$ is PEP in case of Hamming code (7, 4). It is given by Eq. (4.22):

$$\mathrm{PEP}^H = 1 - \left(1 - p^{\mathrm{AC}}\right)\left(1 - p^{\mathrm{HD}}\right)\left(1 - p^{\mathrm{PLH}x}\right) \tag{4.22}$$

For $DM_1$ packet, $\mathrm{PEP}^H$ is given by Eq. (4.23):

$$\mathrm{PEP}^H = 1 - \left(1 - p^{\mathrm{AC}}\right)\left(1 - p^{\mathrm{HD}}\right)\left(1 - p^{\mathrm{PLH1}}\right) \tag{4.23}$$

For $DM_3$ packet, $PEP^H$ is given by Eq. (4.24):

$$PEP^H = 1 - \left(1 - p^{AC}\right)\left(1 - p^{HD}\right)\left(1 - p^{PLH3}\right) \tag{4.24}$$

For $DM_5$ packet, $PEP^H$ is given by Eq. (4.25):

$$PEP^H = 1 - \left(1 - p^{AC}\right)\left(1 - p^{HD}\right)\left(1 - p^{PLH5}\right) \tag{4.25}$$

From the last equations, it is cleared that the PEP depends on the SNR of the channel, where each term in Eq. (4.25) is a function of the SNR. These probabilities of $DM_x$ packets are plotted in Fig. 4.2 on perfect interleaved channel and using Hamming code (7, 4) for $DM_x$ payload encoding. As shown in Fig. 4.2, the performance gain of $DM_x$ packets in case of Hamming (7, 4) code is about 1.5 dB over $DM_x$ packets in standard case. Figures 4.1 and 4.2 reveal that the Hamming code (7, 4) has the better performance for Bluetooth frames if the redundancy length is tolerated. This means that shorter Hamming codes are preferred to longer Hamming codes [8].

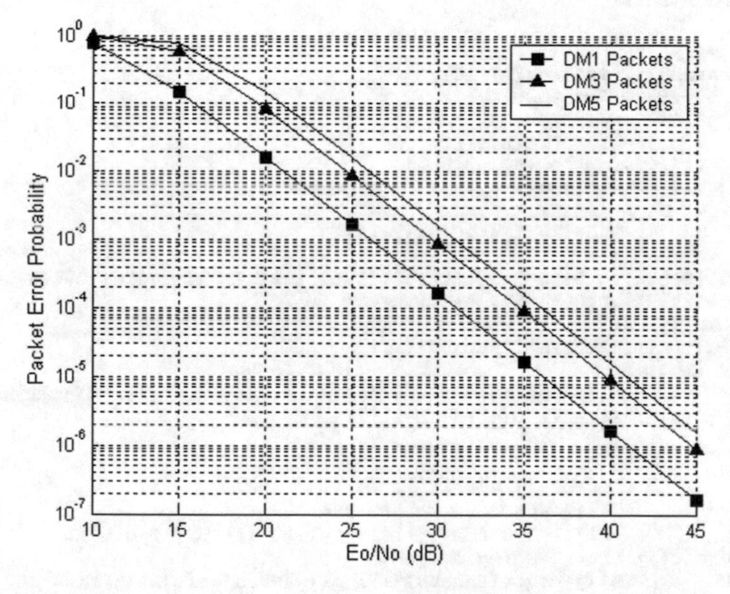


**Fig. 4.2** Analytical PEP of $DM_x$ Hamming (7, 4) (interleaved channel case)

#Matlab Code of Hamming (7, 4)#

```
%==============================IN THE NAME OF ALLAH
====================

% ==============================ANALYTICAL
ANALYSIS======================

%  ====================RELATION BETWEEN THE SNR TO PER =========
======
clc
clear all
close all
k2=1;% input('the length of dataword header part = ');
k3=4;% input('the length of dataword payload part = ');
k1=30; %input('the length of dataword access code part = ');
  n1=64;%input('the length of codeword access code part = ');
  n2=3; %input('the length of codeword header part = ');
  n3=7; %input('the length of codeword payload part = ');

%snr=[10;15;20;25;30;35;40;45;50];
%%%%%%%%%%%%%%%%%%%%%%%%%%%%%%%%%%%%%%%%%%%%%%%%%%%%%%%%%%%%%%%%%%%%%%%
%%%%%%%%%%%%%%%%%%%%%%%%%%%%%%%%%%%%%
  %snr=[0:2:60];
   snr=[0:1:25];
y=length(snr);

for j=1:y
    snrabs(j)=10^(snr(j)/10);
     %####### For AWGN BEP
     %A(j)=sqrt(snrabs(j));
      % p(j)=0.5*erfc(A(j));
      %######for Fading channel
      %A(j)=1/snrabs(j);
      %  B(j)=1+A(j)
      % p(j)=0.5*(1-(1/sqrt(B(j))));

%%%%%%%%%%%%%%%%%%%%%%%%%%%%%%%%%%%%%%%%%%%%%%%%%%%%%%%%%%%%%%%%%%%%%%%
%%%%%%%%%%%%%%%%%%%%%%%%%%%%%%%%%%%%
    pew1(j)=0;
    %p(j)=1/(4*(k1/n1)*snrabs(j));
 %for i1=7:n1
        % pew1(j)=pew1(j)+(factorial(n1)/(factorial(i1)*factorial(n1-
i1)))*((p(j)^i1)*((1-p(j))^(n1-i1)));

            %end
            % pacbad(j)=pew1(j);
            %#####  Uncoded Header Field
             %A1(j)=1/snrabs(j);B1(j)=1+A1(j); p1(j)=0.5*(1-
(1/sqrt(B1(j)))); %Fading channel
             A1(j)=sqrt(snrabs(j)); p1(j)=0.5*erfc(A1(j)); % For AWGN
Channel
            pbadac(j)=(1-p1(j))^30;

%%%%%%%%%%%%%%%%%%%%%%%%%%%%%%%%%%%%%%%%%%%%%%%%%%%%%%%%%%%%%%%%%%%%%%%
%%%%%%%%%%%%%%%%%%%%%%%%%%%%%%%%%%%%%
```

```
          pew2(j)=0;
          %p(j)=1/(4*(k2/n2)*snrabs(j));

      %for i2=2:n2

           %
pew2(j)=pew2(j)+(factorial(n2)/(factorial(i2)*factorial(n2-
i2)))*((p(j)^i2)*((1-p(j))^(n2-i2)));

          % end
          % pbad(j)=1-((1-pew2(j))^18);
          %#####  Uncoded AC Field
          %A2(j)=1/snrabs(j);B2(j)=1+A2(j);p2(j)=0.5*(1-
(1/sqrt(B2(j)))); %Fading channel
          A2(j)=sqrt(snrabs(j)); p2(j)=0.5*erfc(A2(j)); % For AWGN
Channel
         pbadhd(j)=(1-p2(j))^18;

%%%%%%%%%%%%%%%%%%%%%%%%%%%%%%%%%%%%%%%%%%%%%%%%%%%%%%%%%%%%%%%%%%%%%%
%%%%%%%%%%%%%%%%%%%%%%%%%%%%%%%%%%%%%
          pew31(j)=0;
           A3(j)=sqrt((k3/n3)*snrabs(j)); p3(j)=0.5*erfc(A3(j));%
For AWGN Channel
             % A3(j)=1/((k3/n3)*snrabs(j)); B3(j)=1+A3(j);
p3(j)=0.5*(1-(1/sqrt(B3(j))));  %For FAding Channel
     for i3=2:n3


pew31(j)=pew31(j)+(factorial(n3)/(factorial(i3)*factorial(n3-
i3)))*((p3(j)^i3)*((1-p3(j))^(n3-i3)));

     end
        pbaddm1(j)=1-((1-pew31(j))^(16+4));
        pbaddm3(j)=1-((1-pew31(j))^(100+4));
        pbaddm5(j)=1-((1-pew31(j))^(183+4));
    pbaddm11(j)-1-((pbadac(j))*(pbadhd(j))*(1-pbaddm1(j)));
    pbaddm33(j)=1-((pbadac(j))*(pbadhd(j))*(1-pbaddm3(j)));
    pbaddm55(j)=1-((pbadac(j))*(pbadhd(j))*(1-pbaddm5(j)));

        snr;
        DM1(j)=pbaddm11(j);
        DM3(j)=pbaddm33(j);
        DM5(j)=pbaddm55(j);
         %#############  Throughput  Calculationb  #############

        TH_per_dm1(j)=((160+40)*(1-DM1(j)))/((1+1)*625)%
        TH_per_dm3(j)=((1000+40)*(1-DM3(j)))/((3+1)*625)%
        TH_per_dm5(j)=((1830+40)*(1-DM5(j)))/((5+1)*625)%

    end
    %disp(['the amount of ,pbaddm1=:',num2str(pbaddm1)]);
    %disp(['the amount of ,pbaddm3=:',num2str(pbaddm3)]);
    %disp(['the amount of pbad5 =:',num2str(pbaddm5)]);
    %semilogx;
    %semilogy(x);
    hold on;
```

```
%grid on
%snr=[10 15 20 25 30 35 40 45 50];
semilogy(snr,DM1,'b-s');
hold on
semilogy(snr,DM3,'k+:');
hold on
semilogy(snr,DM5,'r+:');
legend('DM1 Packets','DM3 Packets','DM5 Packets');
title('SNR TO PER DMx packets H(7, 4) AWGN');
xlabel('Eo/No (dB)');
ylabel('Packet Error Probability');
grid on

figure(60);

       semilogy(DM1,TH_per_dm1,'b-s');
       hold on
       semilogy(DM3,TH_per_dm3,'k+:');
       hold on
       semilogy(DM5,TH_per_dm5,'r+:');

      legend('THROUGHPUT DM1','THROUGHPUT DM3','THROUGHPUT DM5');
       title('THROUGHPUT to PER of Classic DMx H(7, 4) Packets over
AWGN Channel');
       xlabel('PER');
       ylabel('Throughput Mbps');
       grid on

       figure(70);
       semilogy(snr,TH_per_dm1,'b-s');
       hold on
       semilogy(snr,TH_per_dm3,'k+:');
       hold on
       semilogy(snr,TH_per_dm5,'r+:');
       hold on;
       legend('THROUGHPUT DM1','THROUGHPUT DM3','THROUGHPUT DM5');
       title('THROUGHPUT to SNR of Classic DMx H(7, 4) Packets over
AWGN Channel');
        xlabel('SNR');
       ylabel('Throughput Mbps');
       grid on

       figure(80);

       semilogy(p3,TH_per_dm1,'b-s');
       hold on
       semilogy(p3,TH_per_dm3,'k+:');
       hold on
       semilogy(p3,TH_per_dm5,'r+:');

      legend('THROUGHPUT DM1','THROUGHPUT DM3','THROUGHPUT DM5');
       title('THROUGHPUT to BEP of Classic DMx H(7, 4) Packets over
AWGN Channel');
       xlabel('Bit Error Probability (PEP)');
       ylabel('Throughput Mbps');
       grid on
```

### 4.3.2 BCH Code (15, 7)

BCH code is one of the most powerful block codes. This code will be used for $DM_x$ packet payload only. This is accomplished by dividing the payload into seven-bit segments which are then encoded to 15-bit codeword with BCH code (15, 7). This coding structure will be used for perfectly interleaved Rayleigh fading channel. In this case, the data word and codeword lengths are given as follows [9]:

$n = 15$ bits,

$k = 7$ bits,

$t = 2$ bits.

The previous equations would be modified for the following cases: For the BCH (15, 7) code, the length of the payload will be reduced to keep the Bluetooth frame with the same standard length; this modification makes the following [10]:

**$DM_1$** payload = 112 bit $\rightarrow m = 16$.

**$DM_3$** payload = 700 bit $\rightarrow m = 100$.

**$DM_5$** payload = 1281 bit $\rightarrow m = 183$.

Before writing the new equations, we note that the number of codewords is the same in this case and in case of Hamming (15, 10). The main difference is that the length of payload is reduced by 30% and $t = 2$:

$$P_{\text{Becw}}^{\text{PL}} = \sum_{i=2+1}^{15} \binom{15}{i} P_{\text{b}}^{i} (1 - P_{\text{b}})^{15-i} \tag{4.26}$$

From this equation, we get the codeword error probability, that is, for BCH code (15, 7).

From Eq. (4.3), we get the probability of error in $DM_1$, $DM_3$, and $DM_5$ packets as follows:

**For $DM_1$ Packets**

$DM_1$ packet is one type of $DM_x$ packets. Length of its payload is $\mathbf{L} = 112$ bits. So we note that $m = 16$ codeword. Error probability of encoded payload of $DM_1$ is given by Eq. (4.27):

$$P^{\text{PLB1}} = 1 - \left(1 - P_{\text{Becw}}^{\text{PL}}\right)^{16} \tag{4.27}$$

**$DM_3$ Packets**

$DM_3$ packet is second type of $DM_3$ packets. Length of its payload is $\mathbf{L} = 700$ bits. So we note that $m = 100$ codeword. Error probability of whole $DM_3$ payload is given by Eq. (4.28):

$$P^{\text{PLB3}} = 1 - \left(1 - P_{\text{Becw}}^{\text{PL}}\right)^{100} \tag{4.28}$$

**$DM_5$ Packets**

$DM_5$ packet is third type of $DM_x$ packets. Length of its payload is $\mathbf{L} = 1568$ bits. So we note that $m = 183$ codeword. Error probability of whole $DM_5$ payload is given by Eq. (4.29):

$$P^{\mathrm{PLB5}} = 1 - \left(1 - p_{\mathrm{Becw}}^{\mathrm{PL}}\right)^{183} \tag{4.29}$$

We can conclude PEP$^B$ of DM$_x$ packets by substituting in Eq. (4.6), where PEP$^B$ is Bluetooth PEP in case of BCH (15, 7). It is given by Eq. (4.30):

$$\mathrm{PEP}^B = 1 - \left(1 - p^{\mathrm{AC}}\right)\left(1 - p^{\mathrm{HD}}\right)\left(1 - p^{\mathrm{PLB}x}\right) \tag{4.30}$$

AC and HD error probabilities $P^{\mathrm{AC}}$ and $P^{\mathrm{HD}}$ are given by Eqs. (4.7) and (4.9), respectively.

For DM$_1$ packet, PEP$^B$ is given by Eq. (4.23):

$$\mathrm{PEP}^B = 1 - \left(1 - p^{\mathrm{AC}}\right)\left(1 - p^{\mathrm{HD}}\right)\left(1 - p^{\mathrm{PLB1}}\right) \tag{4.31}$$

For DM$_3$ packet, PEP$^B$ is given by Eq. (4.24):

$$\mathrm{PEP}^B = 1 - \left(1 - p^{\mathrm{AC}}\right)\left(1 - p^{\mathrm{HD}}\right)\left(1 - p^{\mathrm{PLB3}}\right) \tag{4.32}$$

For DM$_5$ packet, PEP$^B$ is given by Eq. (4.25):

$$\mathrm{PEP}^B = 1 - \left(1 - p^{\mathrm{AC}}\right)\left(1 - p^{\mathrm{HD}}\right)\left(1 - p^{\mathrm{PLB5}}\right) \tag{4.33}$$

From the last equations, we conclude that PEP (for BCH (15, 7) is a function of SNR as all terms are functions of SNR. These probabilities (PEP) of DM$_x$ packets are plotted in Fig. 4.3 in case of interleaved channel and using BCH (15, 7) code with DM$_x$ payload [11].

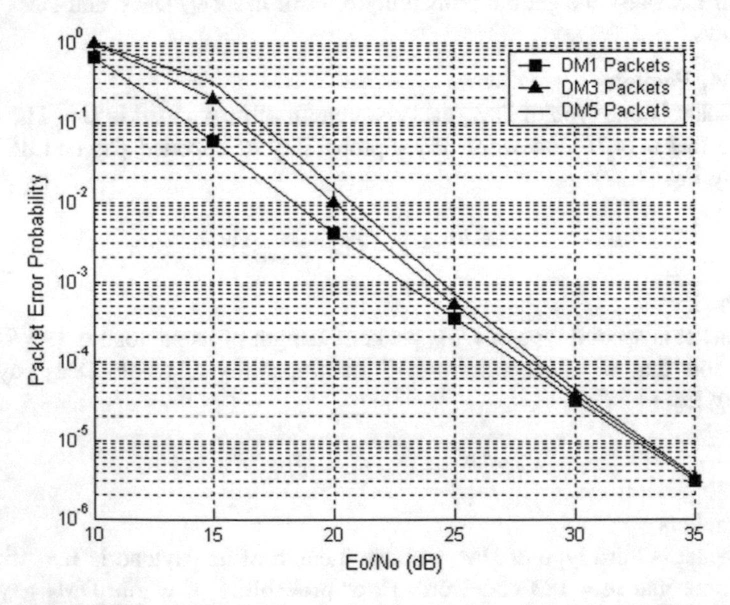


**Fig. 4.3** Analytical PEP of DM$_x$ BCH code (interleaved channel) keeping the Bluetooth frame with the same standard length

PEP of $DM_x$ packets with using BCH (15, 7) code is shown in Fig. 4.3. $DM_x$ packets perform better using this code compared with the $DM_x$ packets' performance in standard case. Figures 4.1, 4.2, and 4.3 reveal that the BCH code has the better performance for Bluetooth frames if the redundancy length is neglected. This means that BCH code is preferred to Hamming codes [12].

```
%==============================IN THE NAME OF ALLAH
===================

% ==============================ANALYTICAL
ANALYSIS=======================

%  ====================RELATION BETWEEN THE SNR TO
PEP================
clc
clear all
close all
k2=1;% input('the length of dataword header part = ');
k3=7;% input('the length of dataword payload part = ');
k1=30; %input('the length of dataword access code part = ');
  n1=64;%input('the length of codeword access code part = ');
  n2=3; %input('the length of codeword header part = ');
  n3=15; %input('the length of codeword payload part = ');

%snr=[10;15;20;25;30;35;40;45;50];
%%%%%%%%%%%%%%%%%%%%%%%%%%%%%%%%%%%%%%%%%%%%%%%%%%%%%%%%%%%%%%%%%%%%%%
%%%%%%%%%%%%%%%%%%%%%%%%%%%%%%%%%%%%
   snr=[0:2:60];
  % snr=[0:1:25];
y=length(snr);

for j=1:y
    snrabs(j)=10^(snr(j)/10);
    %####### For AWGN BEP
     %A(j)=sqrt((k/n)*snrabs(j)); p(j)=0.5*erfc(A(j));% For AWGN
Channel
      %######for Fading channel
     %A(j)=1/snrabs(j);
       % B(j)=1+A(j)
       %p(j)=0.5*(1-(1/sqrt(B(j))));

%%%%%%%%%%%%%%%%%%%%%%%%%%%%%%%%%%%%%%%%%%%%%%%%%%%%%%%%%%%%%%%%%%%%%%
%%%%%%%%%%%%%%%%%%%%%%%%%%%%%%%%%%%
    pew1(j)=0;
        %A1(j)=sqrt((k1/n1)*snrabs(j)); p1(j)=0.5*erfc(A1(j));% For
AWGN Channel
        A1(j)=1/((k1/n1)*snrabs(j)); B1(j)=1+A1(j); p1(j)=0.5*(1-
(1/sqrt(B1(j)))); %For FAding Channel
 for i1=7:n1
         pew1(j)=pew1(j)+(factorial(n1)/(factorial(i1)*factorial(n1-
i1)))*((p1(j)^i1)*((1-p1(j))^(n1-i1)));

            end
             pacbad(j)=pew1(j);

%%%%%%%%%%%%%%%%%%%%%%%%%%%%%%%%%%%%%%%%%%%%%%%%%%%%%%%%%%%%%%%%%%%%%%
%%%%%%%%%%%%%%%%%%%%%%%%%%%%%%%%%%%%
```

```
            pew2(j)=0;
        %A2(j)=sqrt((k2/n2)*snrabs(j)); p2(j)=0.5*erfc(A2(j));% For
AWGN Channel
        A2(j)=1/((k2/n2)*snrabs(j)); B2(j)=1+A2(j); p2(j)=0.5*(1-
(1/sqrt(B2(j)))); %For FAding Channel

       for i2=2:n2


pew2(j)=pew2(j)+(factorial(n2)/(factorial(i2)*factorial(n2-
i2)))*((p2(j)^i2)*((1-p2(j))^(n2-i2)));

            end
            pbad(j)=1-((1-pew2(j))^18);


%%%%%%%%%%%%%%%%%%%%%%%%%%%%%%%%%%%%%%%%%%%%%%%%%%%%%%%%%%%%%%%%%%%%%%%%%
%%%%%%%%%%%%%%%%%%%%%%%%%%%%%%%%%%%%%%%
            pew31(j)=0;

        %A3(j)=sqrt((k3/n3)*snrabs(j)); p3(j)=0.5*erfc(A3(j));% For
AWGN Channel
        A3(j)=1/((k3/n3)*snrabs(j)); B3(j)=1+A3(j); p3(j)=0.5*(1-
(1/sqrt(B3(j)))); %For FAding Channel
      for i3=3:n3


pew31(j)=pew31(j)+(factorial(n3)/(factorial(i3)*factorial(n3-
i3)))*((p3(j)^i3)*((1-p3(j))^(n3-i3)));

      end
         pbaddm1(j)=1-((1-pew31(j))^(2*36));
         pbaddm3(j)=1-((1-pew31(j))^(2*216));
         pbaddm5(j)=1-((1-pew31(j))^(2*393));
     pbaddm11(j)=1-((1-pacbad(j))*(1-pbad(j))*(1-pbaddm1(j)));
     pbaddm33(j)=1-((1-pacbad(j))*(1-pbad(j))*(1-pbaddm3(j)));
     pbaddm55(j)=1-((1-pacbad(j))*(1-pbad(j))*(1-pbaddm5(j)));
     snr;
     DM1(j)=pbaddm11(j);
     DM3(j)=pbaddm33(j);
     DM5(j)=pbaddm55(j);
      %############  Throughput  Calculationb  #############

     TH_per_dm1(j)=((2*108)*(1-DM1(j)))/((1+1)*625);%
     TH_per_dm3(j)=((2*648)*(1-DM3(j)))/((3+1)*625);%
     TH_per_dm5(j)=((2*1179)*(1-DM5(j)))/((5+1)*625);%

end
%disp(['the amount of ,pbaddm1=:',num2str(pbaddm1)]);
%disp(['the amount of ,pbaddm3=:',num2str(pbaddm3)]);
%disp(['the amount of pbad5 =:',num2str(pbaddm5)]);
%semilogx;
%semilogy(x);
hold on;
```

```
%grid on
%snr=[10 15 20 25 30 35 40 45 50];
semilogy(snr,DM1,'b-s');
hold on
semilogy(snr,DM3,'k+:');
hold on
semilogy(snr,DM5,'r+:');
legend('2DM1 Packets','2DM3 Packets','2DM5 Packets');
title('SNR TO PER EDR 2DMx using BCH(15,7)m=3 and t=2 over FADING');
xlabel('Eo/No (dB)');
ylabel('Packet Error Probability');
grid on

figure(60);

       semilogy(DM1,TH_per_dm1,'b-s');
       hold on
       semilogy(DM3,TH_per_dm3,'k+:');
       hold on
       semilogy(DM5,TH_per_dm5,'r+:');

      legend('THROUGHPUT 2DM1','THROUGHPUT 2DM3','THROUGHPUT 2DM5');
       title('THROUGHPUT to PER of EDR 2DMx using BCH(15,7)m=3 and t=2
over FADING');
       xlabel('PER');
       ylabel('Throughput Mbps');
       grid on

       figure(70);
       semilogy(snr,TH_per_dm1,'b-s');
       hold on
       semilogy(snr,TH_per_dm3,'k+:');
       hold on
       semilogy(snr,TH_per_dm5,'r+:');
       hold on;
       legend('THROUGHPUT 2DM1','THROUGHPUT 2DM3','THROUGHPUT 2DM5');
       title('THROUGHPUT to SNR of EDR 2DMx using BCH(15,7)m=3 and t=2
over FADING');
        xlabel('SNR');
       ylabel('Throughput Mbps');
       grid on

       figure(80);
       semilogy(p3,TH_per_dm1,'b-s');
       hold on
       semilogy(p3,TH_per_dm3,'k+:');
       hold on
       semilogy(p3,TH_per_dm5,'r+:');

      legend('THROUGHPUT 2DM1','THROUGHPUT 2DM3','THROUGHPUT 2DM5');
       title('THROUGHPUT to BEP of EDR 2DMx using BCH(15,7)m=3 and t=2
over FADING');
       xlabel('Bit Error Probability (PEP)');
       ylabel('Throughput Mbps');
       grid on
```

## 4.4 Simulation Assumption

In our simulation, we consider a BPSK modulation instead of GFSK modulation, the later which is used in modulation of Bluetooth frames [13].

> In our simulation, we use Monte Carlo simulations to evaluate the performance for Bluetooth packets $DM_x$ and $DH_x$. The results of simulation presented consider the case of a BPSK link over channels. In our experiments we consider a packet of Bluetooth is dropped if there is at least one error after decoding the three portions of Bluetooth frame the AC, the HD, and PL, the last term is payload of encoded Packets DMx [14].
>
> In case of uncoded Bluetooth packets $DH_x$, its packet is discarded if there is single error after decoding the first two portions AC, HD, and detecting errors in uncoded PL after receiving.
>
> In Bluetooth system there are packets types are sent over multi-time slots and others occupy single time slot that is according to the frequency hopping technique. In the presented computer simulation experiments, the frequncy hopping technique is not considered for simplyfing the simulation model. The different number of time slots, 1, 3, and 5 time slots are occupied by $DM_1$, $DM_3$, and $DM_5$ packet respectively, which are occupied by the WPAN-Bluetooth packets is considered. Bluetooth frames have different lengths according to its time-slots which are occupied by these frames. The cannel of simulation is considered time-invariant. That means $\Delta f$ = zero (Doppler spread equal zero) [14].

In our simulation, hard decision is assumed at receiver in decoding process for different channel code situations. In simulation, we neglect interference effects. To analyze the performance of Bluetooth system and get fair comparison between different coding schemes, the length of Bluetooth packet will be constant for each one of simulation cases and that will be realized by changing the length of payload [14].

Payload length for shortened Hamming code (15, 10) before encoding process will be 160 bits, 1000 bits, and 1830 bits for $DM_1$, $DM_3$, and $DM_5$, respectively.

Payload length for Hamming code (7, 4) before encoding process will be 136 bits, 856 bits, and 1568 bits for $DM_1$, $DM_3$, and $DM_5$, respectively. Payload decreased about 14%.

Payload length for BCH code (15, 7) before encoding process will be 112 bits, 700 bits, and 1281 bits for $DM_1$, $DM_3$, and $DM_5$, respectively. Payload deceased by 30%.

Payload length for cyclic code (15, 11) before encoding process will be 165 bits, 1001 bits, and 1837 bits for $DM_1$, $DM_3$, and $DM_5$, respectively. We note the payload lengths are the same payload lengths in standard Bluetooth frame approximately. But length of Bluetooth packets will be less than the standard length that is due to decreasing the length of encoded payloads.

In case of uncoded packet simulation, the maximum allowed data in the PL uncoded packet will be $DH_1$-240 bits, $DH_3$-1500, and $DH_5$-2745 bits with no FEC.

In our simulation, block-fading channel is considered. It is slow and frequency non-selective, where symbols in a block undergo a constant fading effect. The fading coefficients are uncorrelated from block to block regardless of the length of block. That is, it is constant over one hop and independent from hop to hop (block-fading channel) Rayleigh fading statistics are also considered, which well represent the case of non-line-of-sight links [15].

## 4.5 Simulation Results

In this section, there are two sections:

**Firstly**, simulation of basic Bluetooth packet types ($DM_x$, $DH_x$) over AWGN and Rayleigh-flat fading channels.
**Secondly**, simulation of proposed other block codes, which are Hamming code (7, 4), BCH code (15, 7), and cyclic code (15, 11) over AWGN and Rayleigh-flat fading channels.

#Matlab code of DM*x* classic packets simulation#

```
clc
clear all
close all
 %_-_-___________BLUEOOTH SIMULATION of classic with different
channel_________________-_-_-_-%

 %THE LENGTH OF BT_PACKET  BEFORE ENCODING PROCESS
 % LENGTH=4+(24+6)+4+18+PAYLOAD
                                  %MEDIUM RATE(K)
 %SO THE LENGTH OF  ====> DM1=56+PAY(160) =216
===========>FEC(15,10)HAM.
 %SO THE LENGTH OF  ====> DM3=56+PAY(1000)=1056
===========>FEC(15,10)HAM.
 %SO THE LENGTH OF  ====> DM5=56+PAY(1830)=1886
===========>FEC(15,10)HAM.
            %______________HIGH RATE(K)__________________%

 %SO THE LENGTH OF  ====> DH1=56+PAY(240) =296  ===========> NO FEC
 %SO THE LENGTH OF  ====> DH3=56+PAY(1500)=1556 ===========> NO FEC
 %SO THE LENGTH OF  ====> DH5=56+PAY(2745)=2801 ===========> NO FEC
            %_____________AFTER ENCODING____________%
 % THE LENGTH OF BT_PACKET AFTER ENCODING PROCESS
 %LENGTH= 4+64(SYNC WORD)+4+54(HEADER)+PAYLOAD
                                  % MEDIUM RATE(N)
 %SO THE LENGTH OF  ====> DM1=126+PAY(240) =366
<===========FEC(15,10)HAM.
 %SO THE LENGTH OF  ====> DM3=126+PAY(1500)=1626
<===========FEC(15,10)HAM.
 %SO THE LENGTH OF  ====> DM5=126+PAY(2745)=2871
<===========FEC(15,10)HAM.
              %______________HIGH RATE(N)__________________%

 %SO THE LENGTH OF  ====> DH1=126+PAY(240) =366  ===========> NO FEC
 %SO THE LENGTH OF  ====> DH3=126+PAY(1500)=1626 ===========> NO FEC
 %SO THE LENGTH OF  ====> DH5=126+PAY(2745)=2871 ===========> NO FEC
 % THE RATE WILL BE AS SHOWN

            % DM1  -------> R1=K1/N1=216/366.
            % DM3  -------> R3=K3/N3=1056/1626.
            % DM5  -------> R5=K5/N5=1886/2871.
            % DH1----------> RH1=K'1/N'1=296/366.
            % DH3  -------> RH3=K'3/N'3=1556/1626.
            % DH5  -------> RH5=K'5/N'5=2801/2871.
```

```
%HERE WE KNOW RATE OF EACH BLUETOOTH PACKET TRYING TO CHANGE THE
%FEC AND COMPARE 1-FIRST USE BLOCK CODES
%                 2-SECOND USE CONV.ECODER WITH PUNCHERING PROCESS
%                       TO KEEP THE RATE AS MUCH AS WE CAN
%_____________________________________________________________
%***********#########INPUT VARIABLE

    segm1=10 ;% input('the no of k data in of encoder =  ');
%segma1=10
    segm2=15;% input('the no of k data in of decoder =  ');
%segma2=15
    n2=63    ;% input('the no of codeword bch-access  =  ');
%n2=64
    k2=30    ;% input('the no of k data bch encoder =  ');        %
k2=30
    N=input('the NO of iteration                       =  ');
    n1=15;%input('the NO of codeword length  ham(15,11) n1=  ');
    k1=11;%input('the NO of dataword length  ham(15,11) k1=  ');
    t =2;%input('the no of corrected error bch(31,21)  t =  ');
    %G IS GENERATOR MATRIX FOR HAMMING (15,10)ENCODER

    %G=[1 1 0 1 0 1 0 0 0 0 0 0 0 0 0;
       %0 1 1 0 1 0 1 0 0 0 0 0 0 0 0;
       %1 1 1 0 0 0 0 1 0 0 0 0 0 0 0;
       %0 1 1 1 0 0 0 0 1 0 0 0 0 0 0;
       %0 0 1 1 1 0 0 0 0 1 0 0 0 0 0;
       %1 1 0 0 1 0 0 0 0 0 1 0 0 0 0;
       %1 0 1 1 0 0 0 0 0 0 0 1 0 0 0;
       %0 1 0 1 1 0 0 0 0 0 0 0 1 0 0;
       %1 1 1 1 1 0 0 0 0 0 0 0 0 1 0;
       %1 0 1 0 1 0 0 0 0 0 0 0 0 0 1];

      % H  IS PARITY_CHECK MATRIX FOR HAMMING (15,10) DECODER

            %H=[1 0 0 0 0 1 0 1 0 0 1 1 0 1 1;
              % 0 1 0 0 0 1 1 1 1 0 1 0 1 1 0;
               %0 0 1 0 0 0 1 1 1 1 0 1 0 1 1;
               %0 0 0 1 0 1 0 0 1 1 0 1 1 1 0;
               %0 0 0 0 1 0 1 0 0 1 1 0 1 1 1];

          %snr=[0:1:10];
            snr=[0:5:35];
          y=length(snr);
          snrabs=[];
          counter_iter=0;
          for j=1:y
              counter_iter=counter_iter+1
              snrabs(j)=10^(snr(j)/10);
              sigma1=1/sqrt(2*snrabs(j));%     k_1=224    &   n_1=373
      DM_PACKETS
              sigma3=1/sqrt(2*snrabs(j));  %k_3=1064    &  n_1=1613
      DM_PACKETS
              sigma5=1/sqrt(2*snrabs(j));  % k_5=1848    &   n_1=2853
      DM_PACKETS
              error_dh1=0;
              error_dm1=0;
              error_dh3=0;
```

```
error_dm3=0;
error_dh5=0;
error_dm5=0;
 per1=0;
 per2=0;
 per3=0;

 for i=1:N


%p1=round(rand(1,4));                    %preample
p2=round(rand(1,30));                   %sync word LAP=24+6
%p3=round(rand(1,4));                    %trailer
p4=round(rand(1,18));                   %header
dm1=round(rand(1,165));                 %payload dm1 packet fec
(15,10)ham
dm3=round(rand(1,1001));                %payload dm3 packet ,,,,,,,,,,,
dm5=round(rand(1,1837));                %payload dm5 packet,,,,,,,,,,,
%-----------------------------------------
%dh1=round(rand(1,240));                %payload dh1 packet  no fec
%dh3=round(rand(1,1500));                %payload dh3 packet ,,,,,
%dh5=round(rand(1,2745));                %payload dh5 packet  ,,,,,
%################################################################
bt_pkt1=[p2,p4,dm1];
bt_pkt3=[p2,p4,dm3];
bt_pkt5=[p2,p4,dm5];

%----------------------------------=



%bt_pkth1=[p1,p2,p3,p4,dh1];
%bt_pkth3=[p1,p2,p3,p4,dh3];
%bt_pkth5=[p1,p2,p3,p4,dh5];
%**************************************************************
%----------------------------------


% p2 sync word use code bch code(64,30)

%p2=gf(p2);              %BCH CODE ACCESS CODE PART IN GENERAL FORM
%[genpoly,t]=bchgenpoly(n2,k2);
 %pg = bchpoly(n2, k2);
 p2_encode=bchenco(p2,n2,k2);

%***************************************
% P4 IS HEADER USE REPETITION CODE 1/3
for k=1:18
    for l=1:3
        p4_encode(3*(k-1)+l)=p4(k);
    end
end
```

```
    %************************************************
    %P5=PDM OR PDH OR OTHER DM USE HAMM.CODE (15,10)
    % BUT DH NO FEC
    dm1_encode=encode(dm1,n1,k1,'hamming/binary');
%ENCODING(X,G,SEGM) IT IS FUNCTION

    dm3_encode=encode(dm3,n1,k1,'hamming/binary');

    dm5_encode=encode(dm5,n1,k1,'hamming/binary');

    %******************************************************
    %so the bt_pkt after encoding & before sending  it as will be
    %shown

    bt_pkt_enc1=[p2_encode,p4_encode,dm1_encode'];
    bt_pkt_enc3=[p2_encode,p4_encode,dm3_encode'];
    bt_pkt_enc5=[p2_encode,p4_encode,dm5_encode'];
    %------------------------------------
    %bt_pkt_ench1=[p1,p2_encode,p3,p4_encode,dh1];
    %bt_pkt_ench3=[p1,p2_encode,p3,p4_encode,dh3];
    %bt_pkt_ench5=[p1,p2_encode,p3,p4_encode,dh5];

    %******************MOD. AS BPSK FOR ALL PACKET TYPE*****
    %***************** MODULATION
    mod_bt_pkt1=2*bt_pkt_enc1-1;
    mod_bt_pkt3=2*bt_pkt_enc3-1;
    mod_bt_pkt5=2*bt_pkt_enc5-1;

    %------------------------------

    %mod_bt_pkth1=2*bt_pkt_ench1-1;
    %mod_bt_pkth3=2*bt_pkt_ench3-1;
    %mod_bt_pkth5=2*bt_pkt_ench5-1;

%###################################################################

%###################################################################
    %######################### 1- AWGN Channel
############################
    %rec_bt_pkt1=mod_bt_pkt1+sigma1*randn(1,length(bt_pkt_enc1));
    %rec_bt_pkt3=mod_bt_pkt3+sigma1*randn(1,length(bt_pkt_enc3));
    %rec_bt_pkt5=mod_bt_pkt5+sigma1*randn(1,length(bt_pkt_enc5));

%###################################################################

    %**##################**2-SLOW FADING  ##################

    %deep_fad=sqrt(0.5*(randn(1,1)).^2+0.5*(randn(1,1)).^2);
   % ##############
   % bt_pkt_d_fad1=mod_bt_pkt1*deep_fad;
    %rec_bt_pkt1=bt_pkt_d_fad1+sigma1*randn(1,length(bt_pkt_enc1));
    %##################
```

```
    %bt_pkt_d_fad3=mod_bt_pkt3*deep_fad;
    %rec_bt_pkt3=bt_pkt_d_fad3+sigma3*randn(1,length(bt_pkt_enc3));
    %###################
    %bt_pkt_d_fad5=mod_bt_pkt5*deep_fad;
    %rec_bt_pkt5=bt_pkt_d_fad5+sigma5*randn(1,length(bt_pkt_enc5));


%###################################################################
    %############### 3- FLAT FAST BLOCK FADING CHANEL
###############
                              % STANDARD ENCODED PACKETS
   %
prf_fad1=sqrt((0.5*(randn(1,length(bt_pkt_enc1))).^2)+(0.5*(randn(1,le
ngth(bt_pkt_enc1))).^2));
   % bt_pkt_p_fad1=mod_bt_pkt1.*prf_fad1;
   % rec_bt_pkt1=bt_pkt_p_fad1+sigma1*randn(1,length(bt_pkt_enc1));


%prf_fad3=sqrt((0.5*(randn(1,length(bt_pkt_enc3))).^2)+(0.5*(randn(1,l
ength(bt_pkt_enc3))).^2));
    %bt_pkt_p_fad3=mod_bt_pkt3.*prf_fad3;
    %rec_bt_pkt3=bt_pkt_p_fad3+sigma3*randn(1,length(bt_pkt_enc3));

   %
prf_fad5=sqrt((0.5*(randn(1,length(bt_pkt_enc5))).^2)+(0.5*(randn(1,le
ngth(bt_pkt_enc5))).^2));
   % bt_pkt_p_fad5=mod_bt_pkt5.*prf_fad5;
   % rec_bt_pkt5=bt_pkt_p_fad5+sigma5*randn(1,length(bt_pkt_enc5));

%###################################################################

%###################################################################
    %###############    4- CORRELATED FADING     ###############
                              % STANDARD ENCODED PACKETS
     [row,column]=size(bt_pkt_enc1);
    a=jack_fading(column);
    rec_bt_pkt1=a.*(mod_bt_pkt1)+sigma1*randn(1,length(bt_pkt_enc1));

    [row,column]=size(bt_pkt_enc3);
    a=jack_fading(column);
    rec_bt_pkt3=a.*(mod_bt_pkt3)+sigma3*randn(1,length(bt_pkt_enc3));

    [row,column]=size(bt_pkt_enc5);
    a=jack_fading(column);
    rec_bt_pkt5=a.*(mod_bt_pkt5)+sigma5*randn(1,length(bt_pkt_enc5));

%###################################################################


    %**********************DEMODULATION
```

```
    % demod_bt_pkt1=rec_bt_pkt1>0;
     %_______________
      for r=1:length(rec_bt_pkt1)
         if rec_bt_pkt1(r)>0

    demod_bt_pkt1(r)=1;
else
    demod_bt_pkt1(r)=0;
end
end
    %demod_bt_pkt3=rec_bt_pkt3>0;
    %______________________
      for r=1:length(rec_bt_pkt3)
         if rec_bt_pkt3(r)>0

    demod_bt_pkt3(r)=1;
else
    demod_bt_pkt3(r)=0;
end
end
   % demod_bt_pkt5=rec_bt_pkt5>0;
   %________________________
    for r=1:length(rec_bt_pkt5)
        if rec_bt_pkt5(r)>0

    demod_bt_pkt5(r)=1;
else
    demod_bt_pkt5(r)=0;
end
end

    %*************DM1

    %p1_rec1=demod_bt_pkt1(1,1:4);
    p2_rec1=demod_bt_pkt1(1,1:63);          %IT FOR DH PKT ONE FROM DM
ONE FROM DH
    %p3_rec1=demod_bt_pkt1(1,68:71);
    p4_rec1=demod_bt_pkt1(1,64:117);
    p5_rec1=demod_bt_pkt1(1,118:length(bt_pkt_enc1));

    %##########################

    %p1_dec1=p1_rec1;
    p2_dec1=bchdeco(p2_rec1,k2,6);
    %p3_dec1=p3_rec1;
    %__________________________

     %p4 IS HEADER REPETION CODE
    %_____________________________________
    for k=1:18
        counter=0;
        for l=1:3
```

```
            if p4_rec1(3*(k-1)+l)==1
                counter=counter+1;
            end
        end
        if counter >1
            p4_dec1(k)=1;
        else
            p4_dec1(k)=0;
        end
    end
       %###########################
        p5_dec1=decode(p5_rec1',n1,k1,'hamming/binary');

          %*************DM3

    %p1_rec3=demod_bt_pkt3(1,1:4);
    p2_rec3=demod_bt_pkt3(1,1:63);          %IT FOR DH PKT ONE FROM DM
ONE FROM DH
    %p3_rec3=demod_bt_pkt3(1,68:71);
    p4_rec3=demod_bt_pkt3(1,64:117);
    p5_rec3=demod_bt_pkt3(1,118:length(bt_pkt_enc3));

    %###########################

    %p1_dec3=p1_rec3;
    p2_dec3=bchdeco(p2_rec3,k2,6);
    %p3_dec3=p3_rec3;
    %______________________________

     %p4 IS HEADER REPETION CODE
    %__________________________________________
    for k=1:18
        counter=0;
        for l=1:3
            if p4_rec3(3*(k-1)+l)==1
                counter=counter+1;
            end
        end
        if counter >1
            p4_dec3(k)=1;
        else
            p4_dec3(k)=0;
        end
    end
     %###########################
          p5_dec3=decode(p5_rec3',n1,k1,'hamming/binary');

          %*************DM5

    %p1_rec5=demod_bt_pkt5(1,1:4);
    p2_rec5=demod_bt_pkt5(1,1:63);  %IT FOR DH PKT ONE FROM DM
 ONE FROM DH
```

```
%p3_rec5=demod_bt_pkt5(1,68:71);
p4_rec5=demod_bt_pkt5(1,64:117);
p5_rec5=demod_bt_pkt5(1,118:length(bt_pkt_enc5));

%##########################
    %p1_dec5=p1_rec5;
    p2_dec5=bchdeco(p2_rec5,k2,6);
    %p3_dec5=p3_rec5;
    %____________________________

     %p4 IS HEADER REPETION CODE
    %_______________________________________
    for k=1:18
        counter=0;
        for l=1:3
            if p4_rec5(3*(k-1)+l)==1
                counter=counter+1;
            end
        end
        if counter >1
            p4_dec5(k)=1;
        else
            p4_dec5(k)=0;
        end
    end

     %##########################
          p5_dec5=decode(p5_rec5',n1,k1,'hamming/binary');

%################################################################
%@################################################################
         bt_pkt_dec1=[p2_dec1,p4_dec1,p5_dec1'];
         bt_pkt_dec3=[p2_dec3,p4_dec3,p5_dec3'];
         bt_pkt_dec5=[p2_dec5,p4_dec5,p5_dec5'];
 %################################################################
       x1=0;
       x2=0;
       x3=0;
         x1=sum(xor(bt_pkt_dec1, bt_pkt1));
         error_dm1=error_dm1+x1;
         x2=sum(xor(bt_pkt_dec3, bt_pkt3));
         error_dm3=error_dm3+x2;
         x3=sum(xor(bt_pkt_dec5, bt_pkt5));
         error_dm5=error_dm5+x3;
         if x1>0
             per1=per1+1;
         end
              if x2>0
             per2=per2+1;
         end
          if x3>0
             per3=per3+1;
         end

%################################################################
```

```
%####################################################################
        end
%###################################################################
%###################################################################
        ber_dm1(j)=error_dm1/(length(bt_pkt1)*N);
        ber_dm3(j)=error_dm3/(length(bt_pkt3)*N); %NO OF ITERATION*NO
OF BITS IN PKT
        ber_dm5(j)=error_dm5/(length(bt_pkt5)*N); %NO OF ITERATION*NO
OF BITS IN PKT

        per_dm1(j)=per1/N
        per_dm3(j)=per2/N
        per_dm5(j)=per3/N

        per_Ndm1(j)=per1
        per_Ndm3(j)=per2
        per_Ndm5(j)=per3

        TH_per_dm1(j)=(165*(1-per_dm1(j)))/((1+1)*625)%
        TH_per_dm3(j)=(1001*(1-per_dm3(j)))/((3+1)*625)%
        TH_per_dm5(j)=(1837*(1-per_dm5(j)))/((5+1)*625)%
    end
      figure(1);
       semilogy(snr,ber_dm1,'^-k');
       hold on
       semilogy(snr,ber_dm3,'O-r');
       hold on
       semilogy(snr,ber_dm5,'S-b');
       legend('DM1 Packets[PL=165]','DM3 Packets[PL=1001]','DM5
Packets[PL=1837]');
       title('SNR  TO  BER CLASSIC DMx PACKETSF Correlated FAD v=10m/h
Fc=2.46GHz (Ham-15,11)');
       xlabel('Eb/No');
       ylabel('BER');
       grid on;
       figure(2);
       semilogy(snr,per_dm1,'*-k');
       hold on
       semilogy(snr,per_dm3,'+-r');
       hold on
       semilogy(snr,per_dm5);
       legend('DM1 Packets[PL=165]','DM3 Packets[PL=1001]','2DM5
Packets[PL=1837]');
       title('SNR  TO  PER  CLASSIC DMx PACKETS Correlated FAD v=10m/h
Fc=2.46GHz (Ham-15,11)');
       xlabel('Eb/No');
       ylabel('PER');
       grid on
       figure(3);
       semilogy(snr,per_Ndm1,'*-k');
       hold on
       semilogy(snr,per_Ndm3,'+-r');
       hold on
       semilogy(snr,per_Ndm5);
```

```
        legend('DM1 Packets[PL=165]','DM3 Packets[PL=1001]','DM5
Packets[PL=1837]');
        title('SNR  TO  NPL  CLASSIC DMx PACKETS Correlated FAD v=10m/h
Fc=2.46GHz (Ham-15,11)');
        xlabel('Eb/No');
        ylabel('No OF PACKETS LOSSES (NPL)');
        grid on

         figure(22);
        semilogy(per_dm1,TH_per_dm1,'s-k');
        hold on
        semilogy(per_dm3,TH_per_dm3,'d-k');
        hold on
        semilogy(per_dm5,TH_per_dm5,'^-r')
        legend('DM1 Packets[PL=165]','DM3 Packets[PL=1001]','DM5
Packets[PL=1837]');
        title('PER  TO  Throughput  CLASSIC DMx PACKETS Correlated FAD
v=10m/h Fc=2.46GHz  (Ham-15,11)');
        xlabel('PER');
        ylabel('Throughput Mbps');
        grid on
        figure(33);
        semilogy(snr,TH_per_dm1,'s-k');
        hold on
        semilogy(snr,TH_per_dm3,'d-k');
        hold on
        semilogy(snr,TH_per_dm5,'^-r');
        legend('DM1 Packets[PL=165]','DM3 Packets[PL=1001]','DM5
Packets[PL=1837]');
        title('SNR  TO  Throughput  CLASSIC DMx PACKETS Correlated FAD
v=10m/h Fc=2.46GHz   (Ham-15,11)');
         xlabel('SNR');
        ylabel('Throughput Mbps');
        grid on
```

### *4.5.1 Performance of Basic Bluetooth Packets*

Figures 4.4 and 4.5 show the performance of encoded Bluetooth frames and uncoded frames, respectively over AWGN channel. These figures reveal that performance of encoded frames ($DM_x$) is better than uncoded frames ($DH_x$). Using shortened Hamming code gives code gain about 3 dB. From these figures, it is noted that the $DM_1$ packets and $DH_1$ packets give nearly the same performance gain over $DM_3$ and $DH_3$, respectively. As shown in these results, over the AWGN channel the coding gain of the encoded $DM_x$ frames is not exceeding on 1 dB. That means the uncoded and encoded packets performance over the AWGN channel is very closed.

Both $DM_x$ and $DH_x$ packets in standard form are simulated over AWGN channel, and the results are given in Figs. 4.4 and 4.5, respectively. These figures reveal that the $DM_x$ packets perform better than $DH_x$ packets. Performance gain of $DM_x$ packets over $DH_x$ packets is 2.7–3 dB; this is due to the use of error control code in the payload (PL) of $DM_x$ packets.

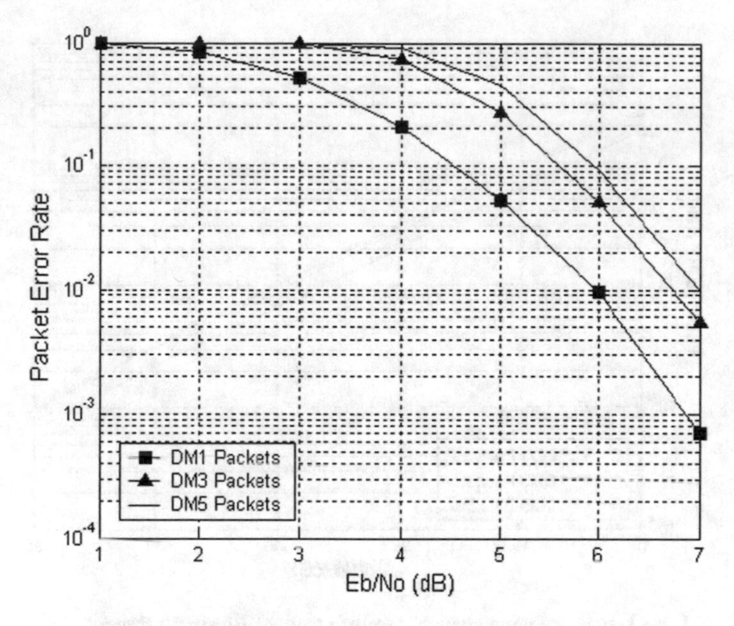


**Fig. 4.4** $DM_x$ packets over AWGN channel case of standard Bluetooth frame

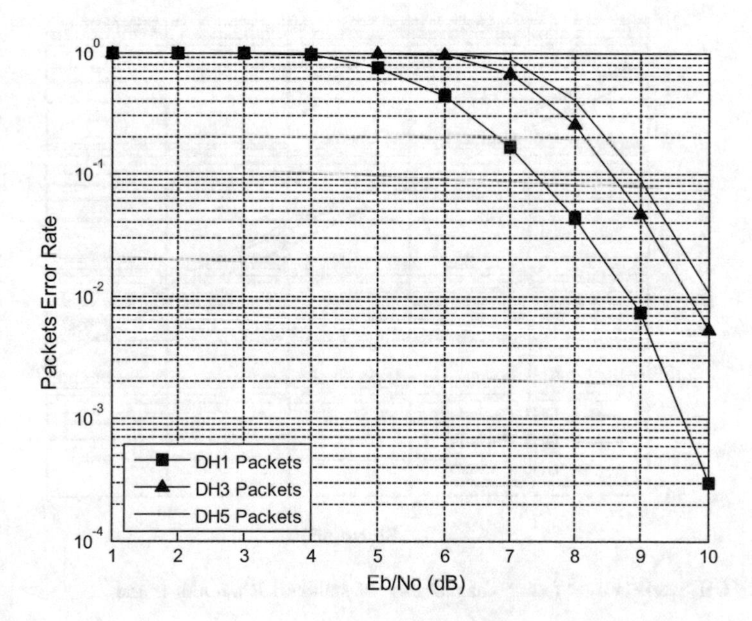


**Fig. 4.5** $DH_x$ packets over AWGN case of standard Bluetooth frames

Figures 4.6 and 4.7 show the performance of encoded Bluetooth frames ($DM_x$) and uncoded frames over Rayleigh-flat fading channel, respectively. Block-fading channel is presented in the computer simulation experiments that is slow and frequency non-selective, where symbols in a block undergo a constant fading effect.

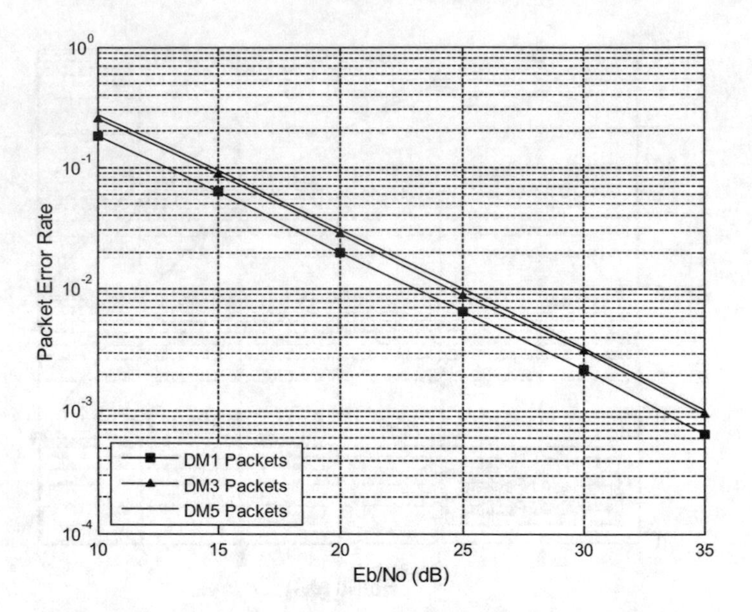


**Fig. 4.6** $DM_x$ packets over fading channel case of standard Bluetooth frame

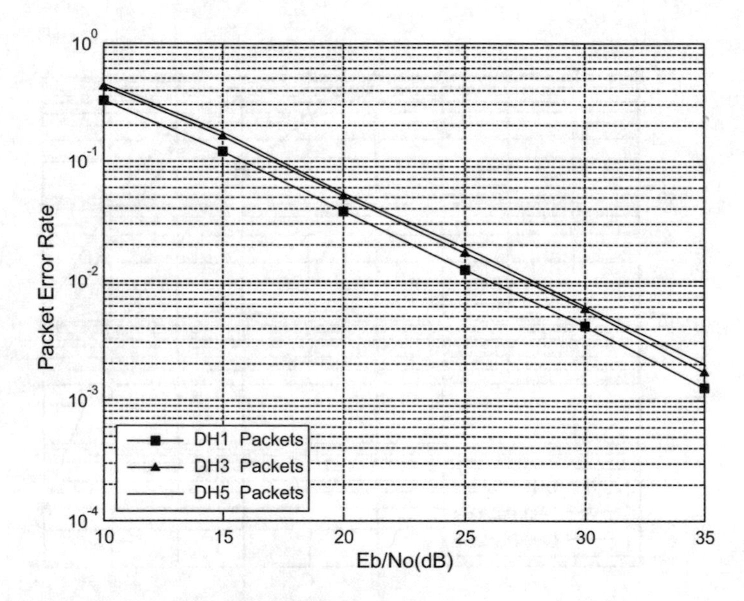


**Fig. 4.7** $DH_x$ packets over fading channel case of standard Bluetooth frame

The WPANs-Bluetooth devices operation do not need to be in line-of-sight requirement [19]. $DM_x$ and $DH_x$ packets in standard form are simulated over Rayleigh-flat fading channel, and the results are given in Figs. 4.6 and 4.7, respectively, which reveal that the $DM_x$ packets perform better than $DH_x$ packets due to using FEC in the payloads of $DM_x$ packets.

Performance gain of $DM_x$ packets over $DH_x$ packets is 2–3 dB, which is due to the use of error control code in the payload (PL) of $DM_x$ packets. As shown in figures, $DM_1$ and $DH_1$ perform better than longer packets $DM_3$ and $DH_3$, respectively. But performance of $DM_3$ and $DH_3$ is closed to $DM_5$ and $DH_5$ performance, respectively.

# Matlab code of classical DH*x* packets simulation [21]#

```
clear all
close all
 %_-_-__________________BLUEOOTH SIMULATION__________________-_-_-_-%

 %THE LENGTH OF BT_PACKET  BEFORE ENCODING PROCESS
 % LENGTH=4+(24+6)+4+18+PAYLOAD
                                  %MEDIUM RATE(K)
 %SO THE LENGTH OF  ====> DM1=56+PAY(160) =216
===========>FEC(15,10)HAM.
 %SO THE LENGTH OF  ====> DM3=56+PAY(1000)=1056
===========>FEC(15,10)HAM.
 %SO THE LENGTH OF  ====> DM5=56+PAY(1830)=1886
===========>FEC(15,10)HAM.
              %______________HIGH RATE(K)__________________%

 %SO THE LENGTH OF  ====> DH1=56+PAY(240) =296  ===========> NO FEC
 %SO THE LENGTH OF  ====> DH3=56+PAY(1500)=1556 ===========> NO FEC
 %SO THE LENGTH OF  ====> DH5=56+PAY(2745)=2801 ===========> NO FEC
              %_____________AFTER ENCODING____________%
 % THE LENGTH OF BT_PACKET AFTER ENCODING PROCESS
 %LENGTH= 4+64(SYNC WORD)+4+54(HEADER)+PAYLOAD
                                   % MEDIUM RATE(N)
 %SO THE LENGTH OF  ====> DM1=126+PAY(240) =366
<===========FEC(15,10)HAM.
 %SO THE LENGTH OF  ====> DM3=126+PAY(1500)=1626
<===========FEC(15,10)HAM.
 %SO THE LENGTH OF  ====> DM5=126+PAY(2745)=2871
<===========FEC(15,10)HAM.
                %_______________HIGH RATE(N)_______________%

 %SO THE LENGTH OF  ====> DH1=126+PAY(240) =366  ===========> NO FEC
 %SO THE LENGTH OF  ====> DH3=126+PAY(1500)=1626 ===========> NO FEC
 %SO THE LENGTH OF  ====> DH5=126+PAY(2745)=2871 ===========> NO FEC
 % THE RATE WILL BE AS SHOWN

             % DM1  -------> R1=K1/N1=216/366.
             % DM3  -------> R3=K3/N3=1056/1626.
             % DM5  -------> R5=K5/N5=1886/2871.
             % DH1---------> RH1=K'1/N'1=296/366.
             % DH3  -------> RH3=K'3/N'3=1556/1626.
             % DH5  -------> RH5=K'5/N'5=2801/2871.

    %HERE WE KNOW RATE OF EACH BLUETOOTH PACKET TRYING TO CHANGE THE
    %FEC AND COMPARE 1-FIRST USE BLOCK CODES
    %                2-SECOND USE CONV.ECODER WITH PUNCHERING PROCESS
    %                      TO KEEP THE RATE AS MUCH AS WE CAN


    %____________________________________________________________
    %***********#########INPUT VARIABLE
```

```
    segm1=10 ;% input('the no of k data in of encoder =  ');
%segma1=10
    segm2=15;% input('the no of k data in of decoder =  ');
%segma2=15
    n2=63   ;% input('the no of codeword bch-access =  ');
%n2=64
    k2=30   ;% input('the no of k data bch encoder =  ');        %
k2=30
    N=input('the NO of iteration                    =  ');
    n1=15;%input('the NO of codeword length  ham(15,11) n1=  ');
    k1=11;%input('the NO of dataword length  ham(15,11) k1=  ');
    t =2;%input('the no of corrected error bch(31,21)  t =  ');
    %G IS GENERATOR MATRIX FOR HAMMING (15,10)ENCODER

    %G=[1 1 0 1 0 1 0 0 0 0 0 0 0 0 0;
       %0 1 1 0 1 0 1 0 0 0 0 0 0 0 0;
       %1 1 1 0 0 0 0 1 0 0 0 0 0 0 0;
       %0 1 1 1 0 0 0 0 1 0 0 0 0 0 0;
       %0 0 1 1 1 0 0 0 0 1 0 0 0 0 0;
       %1 1 0 0 1 0 0 0 0 0 1 0 0 0 0;
       %1 0 1 1 0 0 0 0 0 0 0 1 0 0 0;
       %0 1 0 1 1 0 0 0 0 0 0 0 1 0 0;
       %1 1 1 1 1 0 0 0 0 0 0 0 0 1 0;
       %1 0 1 0 1 0 0 0 0 0 0 0 0 0 1];

      % H  IS PARITY_CHECK MATRIX FOR HAMMING (15,10) DECODER

      %H=[1 0 0 0 0 1 0 1 0 0 1 1 0 1 1;
        % 0 1 0 0 0 1 1 1 1 0 1 0 1 1 0;
         %0 0 1 0 0 0 1 1 1 1 0 1 0 1 1;
         %0 0 0 1 0 1 0 0 1 1 0 1 1 1 0;
         %0 0 0 0 1 0 1 0 0 1 1 0 1 1 1];

   % snr=[0:1:10];
   snr=[0:5:35];
   y=length(snr);
   snrabs=[];
   counter_iter=0;
   for j=1:y
       counter_iter=counter_iter+1
       snrabs(j)=10^(snr(j)/10);
       sigma1=1/sqrt(2*snrabs(j));%    k_1=224   &  n_1=373
DM_PACKETS
       sigma3=1/sqrt(2*snrabs(j));  %k_3=1064   &  n_1=1613
DM_PACKETS
       sigma5=1/sqrt(2*snrabs(j));  % k_5=1848   &  n_1=2853
DM_PACKETS
       error_dh1=0;
       error_dm1=0;
       error_dh3=0;
       error_dm3=0;
       error_dh5=0;
       error_dm5=0;
       per1=0;
       per2=0;
       per3=0;
```

```
    for i=1:N

%p1=round(rand(1,4));                    %preample
p2=round(rand(1,30));                   %sync word LAP=24+6
%p3=round(rand(1,4));                    %trailer
p4=round(rand(1,18));                   %header
dh1=round(rand(1,240));                 %payload dm1 packet fec
(15,10)ham
dh3=round(rand(1,1500));                %payload dm3 packet ,,,,,,,,,,,
dh5=round(rand(1,2744));                %payload dm5 packet,,,,,,,,,,,
%---------------------------------------
%dh1=round(rand(1,240));                %payload dh1 packet  no fec
%dh3=round(rand(1,1500));                %payload dh3 packet ,,,,,
%dh5=round(rand(1,2745));                %payload dh5 packet  ,,,,,
%##############################################################
bt_pkt1=[p2,p4,dh1];
bt_pkt3=[p2,p4,dh3];
bt_pkt5=[p2,p4,dh5];

%----------------------------------


%bt_pkth1=[p1,p2,p3,p4,dh1];
%bt_pkth3=[p1,p2,p3,p4,dh3];
%bt_pkth5=[p1,p2,p3,p4,dh5];
%*************************************************************
%----------------------------------


% p2 sync word use code bch code(64,30)

%p2=gf(p2);             %BCH CODE ACCESS CODE PART IN GENERAL FORM
%[genpoly,t]=bchgenpoly(n2,k2);
%pg = bchpoly(n2, k2);
 p2_encode=bchenco(p2,n2,k2);

%************************************
% P4 IS HEADER USE REPETITION CODE 1/3
for k=1:18
    for l=1:3
        p4_encode(3*(k-1)+l)=p4(k);
    end
end

%*******************************************

%**********************************************
%so the bt_pkt after encoding & before sending  it as will be
%shown


bt_pkt_enc1=[p2_encode,p4_encode,dh1];
bt_pkt_enc3=[p2_encode,p4_encode,dh3];
bt_pkt_enc5=[p2_encode,p4_encode,dh5];
%-----------------------------------
%bt_pkt_ench1=[p1,p2_encode,p3,p4_encode,dh1];
%bt_pkt_ench3=[p1,p2_encode,p3,p4_encode,dh3];
%bt_pkt_ench5=[p1,p2_encode,p3,p4_encode,dh5];
```

```
%******************MOD. AS BPSK FOR ALL PACKET TYPE*****
%***************** MODULATION

mod_bt_pkt1=2*bt_pkt_enc1-1;
mod_bt_pkt3=2*bt_pkt_enc3-1;
mod_bt_pkt5=2*bt_pkt_enc5-1;

%------------------------------

%mod_bt_pkth1=2*bt_pkt_ench1-1;
%mod_bt_pkth3=2*bt_pkt_ench3-1;
%mod_bt_pkth5=2*bt_pkt_ench5-1;

%##################################################################

%##################################################################
%#########################  1- AWGN Channel
#############################
%rec_bt_pkt1=mod_bt_pkt1+sigma1*randn(1,length(bt_pkt_enc1));
%rec_bt_pkt3=mod_bt_pkt3+sigma1*randn(1,length(bt_pkt_enc3));
%rec_bt_pkt5=mod_bt_pkt5+sigma1*randn(1,length(bt_pkt_enc5));

%##################################################################

%**###################**2-SLOW FADING  ###################

% deep_fad=sqrt(0.5*(randn(1,1)).^2+0.5*(randn(1,1)).^2);
% ##############
%bt_pkt_d_fad1=mod_bt_pkt1*deep_fad;
%rec_bt_pkt1=bt_pkt_d_fad1+sigma1*randn(1,length(bt_pkt_enc1));
%##################
%bt_pkt_d_fad3=mod_bt_pkt3*deep_fad;
%rec_bt_pkt3=bt_pkt_d_fad3+sigma3*randn(1,length(bt_pkt_enc3));
%###################
%bt_pkt_d_fad5=mod_bt_pkt5*deep_fad;
%rec_bt_pkt5=bt_pkt_d_fad5+sigma5*randn(1,length(bt_pkt_enc5));

%##################################################################
%################ 3- FLAT FAST BLOCK FADING CHANEL
################
                     % STANDARD ENCODED PACKETS
%
prf_fad1=sqrt((0.5*(randn(1,length(bt_pkt_enc1))).^2)+(0.5*(randn(1,le
ngth(bt_pkt_enc1))).^2));
% bt_pkt_p_fad1=mod_bt_pkt1.*prf_fad1;
% rec_bt_pkt1=bt_pkt_p_fad1+sigma1*randn(1,length(bt_pkt_enc1));


%prf_fad3=sqrt((0.5*(randn(1,length(bt_pkt_enc3))).^2)+(0.5*(randn(1,l
ength(bt_pkt_enc3))).^2));
%bt_pkt_p_fad3=mod_bt_pkt3.*prf_fad3;
%rec_bt_pkt3=bt_pkt_p_fad3+sigma3*randn(1,length(bt_pkt_enc3));
```

```
    %
prf_fad5=sqrt((0.5*(randn(1,length(bt_pkt_enc5))).^2)+(0.5*(randn(1,le
ngth(bt_pkt_enc5))).^2));
    % bt_pkt_p_fad5=mod_bt_pkt5.*prf_fad5;
    % rec_bt_pkt5=bt_pkt_p_fad5+sigma5*randn(1,length(bt_pkt_enc5));

%####################################################################

%####################################################################

    %################  4- CORRELATED FADING       ################
                        % STANDARD ENCODED PACKETS
    [row,column]=size(bt_pkt_enc1);
    a=jack_fading(column);
    rec_bt_pkt1=a.*(mod_bt_pkt1)+sigma1*randn(1,length(bt_pkt_enc1));

    [row,column]=size(bt_pkt_enc3);
    a=jack_fading(column);
    rec_bt_pkt3=a.*(mod_bt_pkt3)+sigma3*randn(1,length(bt_pkt_enc3));

    [row,column]=size(bt_pkt_enc5);
    a=jack_fading(column);
    rec_bt_pkt5=a.*(mod_bt_pkt5)+sigma5*randn(1,length(bt_pkt_enc5));

%####################################################################


    %**********************DEMODULATION

   % demod_bt_pkt1=rec_bt_pkt1>0;
    %_________________
     for r=1:length(rec_bt_pkt1)
        if rec_bt_pkt1(r)>0

    demod_bt_pkt1(r)=1;
else
    demod_bt_pkt1(r)=0;
end
end
    %demod_bt_pkt3=rec_bt_pkt3>0;
    %________________________
     for r=1:length(rec_bt_pkt3)
        if rec_bt_pkt3(r)>0

    demod_bt_pkt3(r)=1;
else
    demod_bt_pkt3(r)=0;
end
end
   % demod_bt_pkt5=rec_bt_pkt5>0;
   %__________________________
    for r=1:length(rec_bt_pkt5)
        if rec_bt_pkt5(r)>0
```

```
    demod_bt_pkt5(r)=1;
else
    demod_bt_pkt5(r)=0;
end
end


    %*************DM1

    %p1_rec1=demod_bt_pkt1(1,1:4);
    p2_rec1=demod_bt_pkt1(1,1:63);      %IT FOR DH PKT ONE FROM DM
ONE FROM DH
    %p3_rec1=demod_bt_pkt1(1,68:71);
    p4_rec1=demod_bt_pkt1(1,64:117);
    p5_rec1=demod_bt_pkt1(1,118:length(bt_pkt_enc1));

    %###########################

    %p1_dec1=p1_rec1;
    p2_dec1=bchdeco(p2_rec1,k2,6);
    %p3_dec1=p3_rec1;
    %______________________________

     %p4 IS HEADER REPETION CODE
    %________________________________________
    for k=1:18
        counter=0;
        for l=1:3
            if p4_rec1(3*(k-1)+l)==1
                counter=counter+1;
            end
        end
        if counter >1
            p4_dec1(k)=1;
        else
            p4_dec1(k)=0;
        end
    end

         %###########################
          p5_dec1=p5_rec1;

           %*************DM3


    %p1_rec3=demod_bt_pkt3(1,1:4);
    p2_rec3=demod_bt_pkt3(1,1:63);      %IT FOR DH PKT ONE FROM DM
 ONE FROM DH
    %p3_rec3=demod_bt_pkt3(1,68:71);
    p4_rec3=demod_bt_pkt3(1,64:117);
    p5_rec3=demod_bt_pkt3(1,118:length(bt_pkt_enc3));

    %###########################
```

```
%p1_dec3=p1_rec3;
p2_dec3=bchdeco(p2_rec3,k2,6);
%p3_dec3=p3_rec3;
%______________________________________________

 %p4 IS HEADER REPETION CODE
%_________________________________________________________________
for k=1:18
    counter=0;
    for l=1:3
        if p4_rec3(3*(k-1)+l)==1
            counter=counter+1;
        end
    end
    if counter >1
        p4_dec3(k)=1;
    else
        p4_dec3(k)=0;
    end
end

     %##########################
          p5_dec3=p5_rec3;

          %*************DM5

    %p1_rec5=demod_bt_pkt5(1,1:4);
    p2_rec5=demod_bt_pkt5(1,1:63);      %IT FOR DH PKT ONE FROM DM
ONE FROM DH
    %p3_rec5=demod_bt_pkt5(1,68:71);
    p4_rec5=demod_bt_pkt5(1,64:117);
    p5_rec5=demod_bt_pkt5(1,118:length(bt_pkt_enc5));

    %##########################

    %p1_dec5=p1_rec5;
    p2_dec5=bchdeco(p2_rec5,k2,6);
    %p3_dec5=p3_rec5;
    %______________________________________________

     %p4 IS HEADER REPETION CODE
    %_________________________________________________________________
    for k=1:18
        counter=0;
        for l=1:3
            if p4_rec5(3*(k-1)+l)==1
                counter=counter+1;
            end
        end
        if counter >1
            p4_dec5(k)=1;
        else
            p4_dec5(k)=0;
        end
    end
```

```
    %##########################
        p5_dec5=p5_rec5;

 %############################################################
 %@############################################################
        bt_pkt_dec1=[p2_dec1,p4_dec1,p5_dec1];
        bt_pkt_dec3=[p2_dec3,p4_dec3,p5_dec3];
        bt_pkt_dec5=[p2_dec5,p4_dec5,p5_dec5];
  %##############################################################
      x1=0;
      x2=0;
      x3=0;
        x1=sum(xor(bt_pkt_dec1, bt_pkt1));
        error_dm1=error_dm1+x1;
        x2=sum(xor(bt_pkt_dec3, bt_pkt3));
        error_dm3=error_dm3+x2;
        x3=sum(xor(bt_pkt_dec5, bt_pkt5));
        error_dm5=error_dm5+x3;
        if x1>0
            per1=per1+1;
        end
             if x2>0
            per2=per2+1;
        end
         if x3>0

            per3=per3+1;
        end
 %###########################################################

 %###########################################################

     end

 %###########################################################

 %###########################################################
       ber_dm1(j)=error_dm1/(length(bt_pkt1)*N);
       ber_dm3(j)=error_dm3/(length(bt_pkt3)*N); %NO OF ITERATION*NO
OF BITS IN PKT
       ber_dm5(j)=error_dm5/(length(bt_pkt5)*N); %NO OF ITERATION*NO
OF BITS IN PKT

       per_dm1(j)=per1/N
       per_dm3(j)=per2/N
       per_dm5(j)=per3/N

       per_Ndm1(j)=per1
       per_Ndm3(j)=per2
       per_Ndm5(j)=per3

       TH_per_dm1(j)=(240*(1-per_dm1(j)))/((1+1)*625)%
       TH_per_dm3(j)=(1500*(1-per_dm3(j)))/((3+1)*625)%
       TH_per_dm5(j)=(2744*(1-per_dm5(j)))/((5+1)*625)%

   end
```

```
figure(1);
 semilogy(snr,ber_dm1,'^-k');
 hold on
 semilogy(snr,ber_dm3,'O-r');
 hold on
 semilogy(snr,ber_dm5,'S-b');
 legend('DH1 Packets[PL=240]','DH3 Packets[PL=1500]','DH5
Packets[PL=2744]');
 title('SNR  TO  BER CLASSIC 2DHx PACKETSF Correlated FAD
v=10b/h Fc=2.46GHz ');
 xlabel('Eb/No');
 ylabel('BER');
 grid on;
 figure(2);
 semilogy(snr,per_dm1,'*-k');
 hold on
 semilogy(snr,per_dm3,'+-r');
 hold on
 semilogy(snr,per_dm5);
 legend('DH1 Packets[PL=240]','DH3 Packets[PL=1500]','DH5
Packets[PL=2744]');
 title('SNR  TO  PER  CLASSIC DHx PACKETS Correlated FAD v=10b/h
Fc=2.46GHz ');
 xlabel('Eb/No');
 ylabel('PER');
 grid on
 figure(3);
 semilogy(snr,per_Ndm1,'*-k');
 hold on
 semilogy(snr,per_Ndm3,'+-r');
 hold on
 semilogy(snr,per_Ndm5);
 legend('DH1 Packets[PL=240]','DH3 Packets[PL=1500]','DH5
Packets[PL=2744]');
 title('SNR  TO  NPL  CLASSIC DHx PACKETS Correlated FAD v=10b/h
Fc=2.46GHz');
 xlabel('Eb/No');
 ylabel('No OF PACKETS LOSSES (NPL)');
 grid on

  figure(22);
 semilogy(per_dm1,TH_per_dm1,'s-k');
 hold on
 semilogy(per_dm3,TH_per_dm3,'d-k');
 hold on
 semilogy(per_dm5,TH_per_dm5,'^-r')
 legend('DH1 Packets[PL=240]','DH3 Packets[PL=1500]','DH5
Packets[PL=2744]');
 title('PER  TO  Throughput  CLASSIC DHx PACKETS Correlated FAD
v=10b/h Fc=2.46GHz');
 xlabel('PER');
 ylabel('Throughput Mbps');
 grid on
 figure(33);
 semilogy(snr,TH_per_dm1,'s-k');
 hold on
```

```
        semilogy(snr,TH_per_dm3,'d-k');
        hold on
        semilogy(snr,TH_per_dm5,'^-r');
        legend('DH1 Packets[PL=240]','DH3 Packets[PL=1500]','DH5
Packets[PL=2744]');
        title('SNR  TO  Throughput  CLASSIC DHx PACKETS Correlated FAD
v=10b/h Fc=2.46GHz ');
         xlabel('SNR');
        ylabel('Throughput Mbps');
        grid on
```

### 4.5.2 Effects of Proposed Channel Codes on Bluetooth Performance

This section investigates the performance of $DM_x$ packets and $DH_x$ packets over Rayleigh fading and AWGN channels in the case of the proposed coding schemes and compares it to the performance of previous section channel coding schemes in Bluetooth systems. Several experiments are conducted for this purpose.

In the first, both Hamming (7, 4) and BCH (15, 7) coding schemes investigated for an AWGN channel and the results are given in Figs. 4.8 and 4.9, respectively, which reveal the superiority of the BCH (15, 7) code for the AWGN channel. The performance of BCH code is better than the Hamming code (7, 4) and Hamming code (15, 10). BCH code introduces about 1.5 dB performance gains over

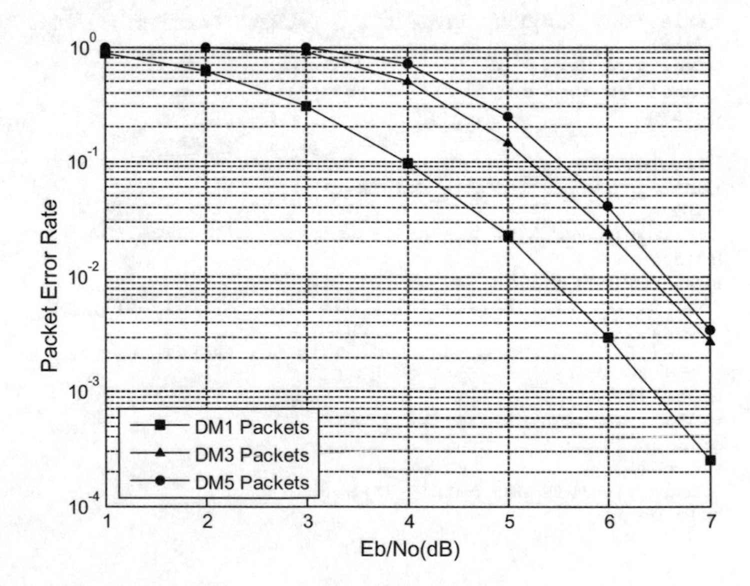


**Fig. 4.8** $DM_x$ packets in case of AWGN channel for Hamming (7, 4) code

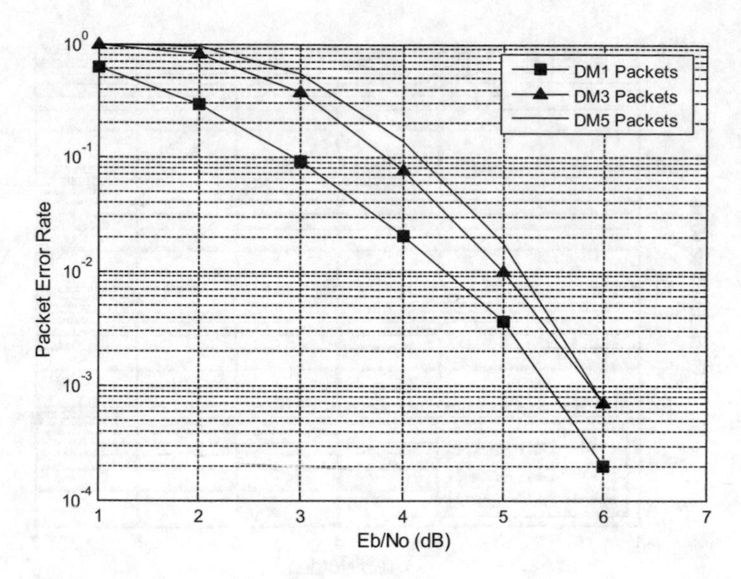


**Fig. 4.9** $DM_x$ packets in case of AWGN channel for BCH (15, 7) code, PL only BCH code (15, 7) PL only

Hamming code (7, 4). In the same experiment, Hamming code (7, 4) performs better than shortened Hamming code (15, 10) code, where it presents performance gain about 0.3 dB over shortened Hamming code (15, 10). This means that shorter Hamming codes are preferred to longer Hamming codes and the best code is the BCH (15, 7) for AWGN channel if the redundancy length is tolerated.

Figure 4.10 shows the simulation result in case of using cyclic code (15, 11) in $DM_x$ packet payloads. This previous simulation experiment reveals that cyclic (15, 11) and Hamming (15, 10) codes give very close performance of $DM_x$ packets over AWGN channel with lower redundancy length.

Several simulation experiments are carried out to investigate the effects of using proposed coding schemes on the $DM_x$ performance compared with $DH_x$ performance. The Hamming (7, 4) and BCH (15, 7) codes are studied for a Rayleigh-flat fading channel. The results are given in Figs. 4.11 and 4.12, respectively.

The result of the computer simulation experiments of utilizing Hamming code (7, 4) and Hamming code (15, 10) for payload encoding are shown in Figs. 4.6 and 4.11 are very close, which means that the Hamming code (15, 10) is better due to its low redundancy bits. Also, the performance of the encoded packets using BCH (15, 7) code over Rayleigh-flat fading channel is investigated. The results are given in Figs. 4.11, 4.12, and 4.13. These results reveal that the BCH code has the best performance for Bluetooth frames if the redundancy length is tolerated.

Simulation result in case of using cyclic code (15, 11) in $DM_x$ packet payloads over Rayleigh-flat fading channel is shown in Fig. 4.13. This figure reveals that cyclic (15, 11) and Hamming (15, 10) codes give similar performance of $DM_x$ packets over with lower redundancy length.

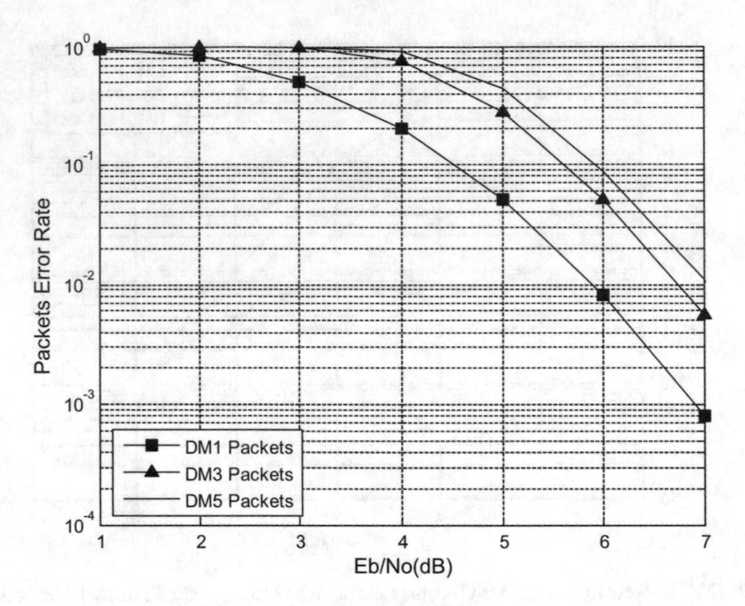


**Fig. 4.10** $DM_x$ packets in case of AWGN channel for cyclic code (15, 11) PL only

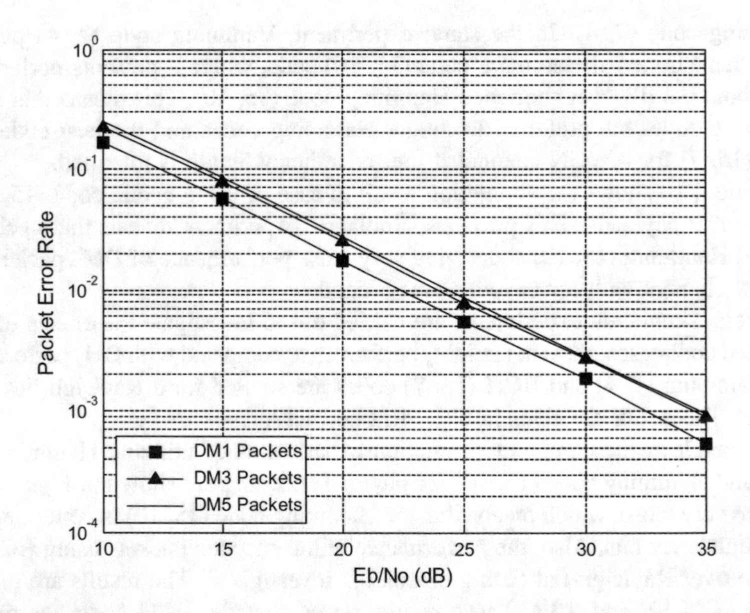


**Fig. 4.11** $DM_x$ packets over fading channel in case of Hamming code (7, 4) for PL only

The result of $DH_x$ performance over Rayleigh-flat fading channel, which is shown in Figs. 4.5 and 4.7, compared to the previous results shows that the $DH_x$ packets are the best choice at high SNR because there is no redundancy bits.

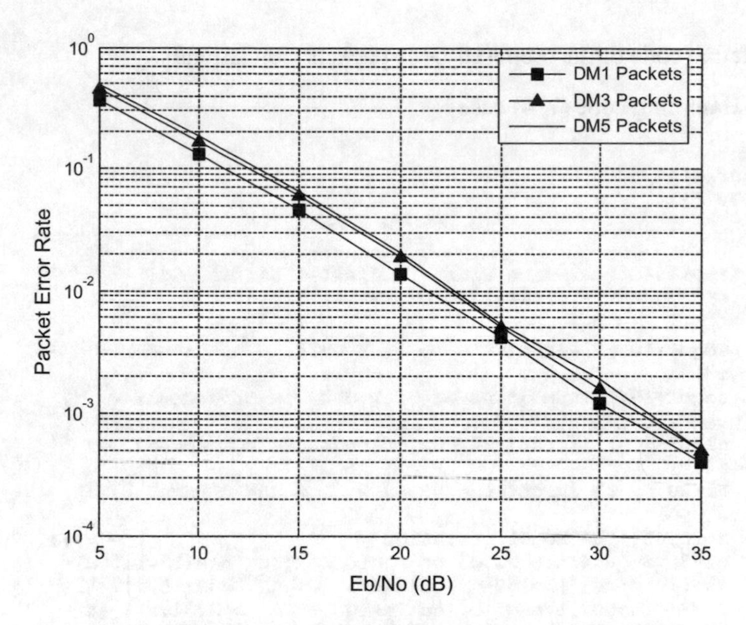


**Fig. 4.12** $DM_x$ packets over fading channel in case of BCH code (15, 7) for PL only

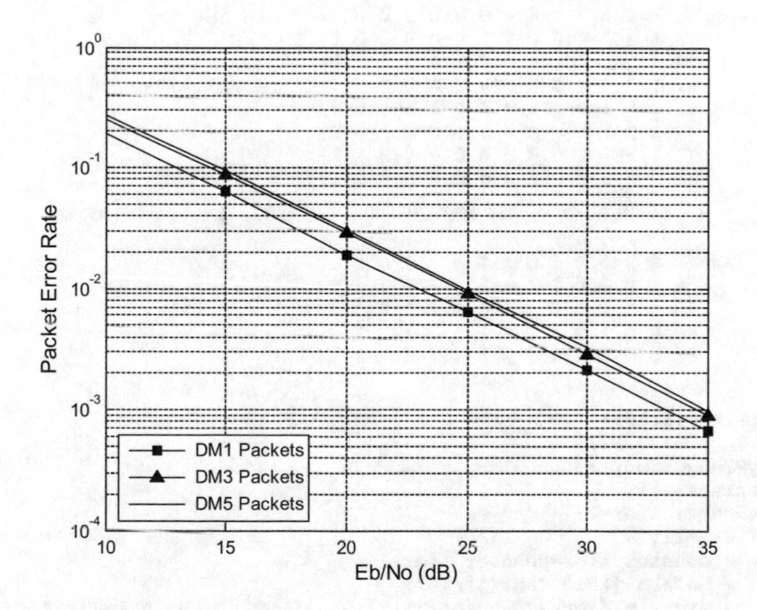


**Fig. 4.13** $DM_x$ packets over fading channel in case of cyclic code (15, 11) for PL only

#Matlab code of the proposed DM*x* packets simulation#

Hamming code (7, 4) case

```
clc
clear all
close all
 %_-_-____________________BLUEOOTH SIMULATION____________________-_-_-_-%

 %&&&&&&&&&&&&&Hamming Code (7, 4)&&&&&&&&&&&&&&&&&
    %***********#########INPUT VARIABLE

    segm1=10 ;% input('the no of k data in of encoder =  ');
%segma1=10
    segm2=15;% input('the no of k data in of decoder =  ');
%segma2=15
    n2=63    ;% input('the no of codeword bch-access  =  ');
%n2=64
    k2=30    ;% input('the no of k data bch encoder =  ');       %
k2=30
    N=input('the NO of iteration                        =  ');
    n1=7;%input('the NO of codeword length  ham(15,11) n1=  ');
    k1=4;%input('the NO of dataword length  ham(15,11) k1=  ');
    t =2;%input('the no of corrected error bch(31,21)  t =  ');
    %G IS GENERATOR MATRIX FOR HAMMING (15,10)ENCODER

    %G=[1 1 0 1 0 1 0 0 0 0 0 0 0 0 0;
       %0 1 1 0 1 0 1 0 0 0 0 0 0 0 0;
       %1 1 1 0 0 0 0 1 0 0 0 0 0 0 0;
       %0 1 1 1 0 0 0 0 1 0 0 0 0 0 0;
       %0 0 1 1 1 0 0 0 0 1 0 0 0 0 0;
       %1 1 0 0 1 0 0 0 0 0 1 0 0 0 0;
       %1 0 1 1 0 0 0 0 0 0 0 1 0 0 0;
       %0 1 0 1 1 0 0 0 0 0 0 0 1 0 0;
       %1 1 1 1 1 0 0 0 0 0 0 0 0 1 0;
       %1 0 1 0 1 0 0 0 0 0 0 0 0 0 1];

    % H  IS PARITY_CHECK MATRIX FOR HAMMING (15,10) DECODER

    %H=[1 0 0 0 0 1 0 1 0 0 1 1 0 1 1;
      % 0 1 0 0 0 1 1 1 1 0 1 0 1 1 0;
       %0 0 1 0 0 0 1 1 1 1 0 1 0 1 1;
       %0 0 0 1 0 1 0 0 1 1 0 1 1 1 0;
       %0 0 0 0 1 0 1 0 0 1 1 0 1 1 1];


  snr=[0:1:10];
 % snr=[0:5:35];
  y=length(snr);
  snrabs=[];
  counter_iter=0;
  for j=1:y
      counter_iter=counter_iter+1
      snrabs(j)=10^(snr(j)/10);
      sigma1=1/sqrt(2*snrabs(j));%    k_1=224   &  n_1=373
DM_PACKETS
```

```
      sigma3=1/sqrt(2*snrabs(j));  %k_3=1064   &  n_1=1613
DM_PACKETS
      sigma5=1/sqrt(2*snrabs(j));  % k_5=1848   &  n_1=2853
DM_PACKETS
      error_dh1=0;
      error_dm1=0;
      error_dh3=0;
      error_dm3=0;
      error_dh5=0;
      error_dm5=0;
      per1=0;
      per2=0;
      per3=0;

      for i=1:N

  %p1=round(rand(1,4));                  %preample
  p2=round(rand(1,30));                  %sync word LAP=24+6
  %p3=round(rand(1,4));                  %trailer
  p4=round(rand(1,18));                  %header
  dm1=round(rand(1,136));                %payload dm1 packet fec
(15,10)ham
  dm3=round(rand(1,856));               %payload dm3 packet ,,,,,,,,,,,
  dm5=round(rand(1,1568));               %payload dm5 packet,,,,,,,,,,,
  %----------------------------------------
  %dh1=round(rand(1,240));               %payload dh1 packet  no fec
  %dh3=round(rand(1,1500));               %payload dh3 packet ,,,,,
  %dh5=round(rand(1,2745));               %payload dh5 packet  ,,,,,
  %###############################################################
  bt_pkt1=[p2,p4,dm1];
  bt_pkt3=[p2,p4,dm3];
  bt_pkt5=[p2,p4,dm5];

  %-----------------------------------=
  %bt_pkth1=[p1,p2,p3,p4,dh1];
  %bt_pkth3=[p1,p2,p3,p4,dh3];
  %bt_pkth5=[p1,p2,p3,p4,dh5];
  %***************************************************************
  %-----------------------------------

  % p2 sync word use code bch code(64,30)

  %p2=gf(p2);                 %BCH CODE ACCESS CODE PART IN GENERAL FORM
  %[genpoly,t]=bchgenpoly(n2,k2);
   %pg = bchpoly(n2, k2);
   p2_encode=bchenco(p2,n2,k2);

  %****************************************
  % P4 IS HEADER USE REPETITION CODE 1/3
  for k=1:18
      for l=1:3
          p4_encode(3*(k-1)+l)=p4(k);
      end
  end
```

```
    %**********************************************
    %P5=PDM OR PDH OR OTHER DM USE HAMM.CODE (15,10)
    % BUT DH NO FEC
    dm1_encode=encode(dm1,n1,k1,'hamming/binary');
%ENCODING(X,G,SEGM) IT IS FUNCTION

    dm3_encode=encode(dm3,n1,k1,'hamming/binary');

    dm5_encode=encode(dm5,n1,k1,'hamming/binary');

    %****************************************************
    %so the bt_pkt after encoding & before sending  it as will be
    %shown

    bt_pkt_enc1=[p2_encode,p4_encode,dm1_encode'];
    bt_pkt_enc3=[p2_encode,p4_encode,dm3_encode'];
    bt_pkt_enc5=[p2_encode,p4_encode,dm5_encode'];
    %------------------------------------
    %bt_pkt_ench1=[p1,p2_encode,p3,p4_encode,dh1];
    %bt_pkt_ench3=[p1,p2_encode,p3,p4_encode,dh3];
    %bt_pkt_ench5=[p1,p2_encode,p3,p4_encode,dh5];


    %******************MOD. AS BPSK FOR ALL PACKET TYPE*****
    %***************** MODULATION
    mod_bt_pkt1=2*bt_pkt_enc1-1;
    mod_bt_pkt3=2*bt_pkt_enc3-1;
    mod_bt_pkt5=2*bt_pkt_enc5-1;

    %------------------------------

    %mod_bt_pkth1=2*bt_pkt_ench1-1;
    %mod_bt_pkth3=2*bt_pkt_ench3-1;
    %mod_bt_pkth5=2*bt_pkt_ench5-1;

%##################################################################

%##################################################################
    %########################## 1- AWGN Channel
##############################
    rec_bt_pkt1=mod_bt_pkt1+sigma1*randn(1,length(bt_pkt_enc1));
    rec_bt_pkt3=mod_bt_pkt3+sigma1*randn(1,length(bt_pkt_enc3));
    rec_bt_pkt5=mod_bt_pkt5+sigma1*randn(1,length(bt_pkt_enc5));

%##################################################################

    %**###################**2-SLOW FADING  ##################

    %deep_fad=sqrt(0.5*(randn(1,1)).^2+0.5*(randn(1,1)).^2);
   % ##############
   % bt_pkt_d_fad1=mod_bt_pkt1*deep_fad;
    %rec_bt_pkt1=bt_pkt_d_fad1+sigma1*randn(1,length(bt_pkt_enc1));
    %##################
    %bt_pkt_d_fad3=mod_bt_pkt3*deep_fad;
    %rec_bt_pkt3=bt_pkt_d_fad3+sigma3*randn(1,length(bt_pkt_enc3));
    %###################
```

```
    %bt_pkt_d_fad5=mod_bt_pkt5*deep_fad;
    %rec_bt_pkt5=bt_pkt_d_fad5+sigma5*randn(1,length(bt_pkt_enc5));


%####################################################################
    %################ 3- FLAT FAST BLOCK FADING CHANEL
################
                          % STANDARD ENCODED PACKETS
   %
prf_fad1=sqrt((0.5*(randn(1,length(bt_pkt_enc1))).^2)+(0.5*(randn(1,le
ngth(bt_pkt_enc1))).^2));
   % bt_pkt_p_fad1=mod_bt_pkt1.*prf_fad1;
   % rec_bt_pkt1=bt_pkt_p_fad1+sigma1*randn(1,length(bt_pkt_enc1));


%prf_fad3=sqrt((0.5*(randn(1,length(bt_pkt_enc3))).^2)+(0.5*(randn(1,l
ength(bt_pkt_enc3))).^2));
    %bt_pkt_p_fad3=mod_bt_pkt3.*prf_fad3;
    %rec_bt_pkt3=bt_pkt_p_fad3+sigma3*randn(1,length(bt_pkt_enc3));

   %
prf_fad5=sqrt((0.5*(randn(1,length(bt_pkt_enc5))).^2)+(0.5*(randn(1,le
ngth(bt_pkt_enc5))).^2));
   % bt_pkt_p_fad5=mod_bt_pkt5.*prf_fad5;
   % rec_bt_pkt5=bt_pkt_p_fad5+sigma5*randn(1,length(bt_pkt_enc5));

%####################################################################
%####################################################################
    %################   4- CORRELATED FADING     ################
                          % STANDARD ENCODED PACKETS
    % [row,column]=size(bt_pkt_enc1);
    %a=jack_fading(column);
    %rec_bt_pkt1=a.*(mod_bt_pkt1)+sigma1*randn(1,length(bt_pkt_enc1));

    %[row,column]=size(bt_pkt_enc3);
    %a=jack_fading(column);
    %rec_bt_pkt3=a.*(mod_bt_pkt3)+sigma3*randn(1,length(bt_pkt_enc3));

    %[row,column]=size(bt_pkt_enc5);
    %a=jack_fading(column);
    %rec_bt_pkt5=a.*(mod_bt_pkt5)+sigma5*randn(1,length(bt_pkt_enc5));

%####################################################################

    %**********************DEMODULATION

   % demod_bt_pkt1=rec_bt_pkt1>0;
    %_________________
     for r=1:length(rec_bt_pkt1)
        if rec_bt_pkt1(r)>0

    demod_bt_pkt1(r)=1;
else
    demod_bt_pkt1(r)=0;
end
```

```
end
    %demod_bt_pkt3=rec_bt_pkt3>0;
    %_______________________
     for r=1:length(rec_bt_pkt3)
        if rec_bt_pkt3(r)>0

    demod_bt_pkt3(r)=1;
else
    demod_bt_pkt3(r)=0;
end
end
   % demod_bt_pkt5=rec_bt_pkt5>0;
   %__________________________
    for r=1:length(rec_bt_pkt5)
        if rec_bt_pkt5(r)>0

    demod_bt_pkt5(r)=1;
else
    demod_bt_pkt5(r)=0;
end
end

    %*************DM1

    %p1_rec1=demod_bt_pkt1(1,1:4);
    p2_rec1=demod_bt_pkt1(1,1:63);      %IT FOR DH PKT ONE FROM DM
ONE FROM DH
    %p3_rec1=demod_bt_pkt1(1,68:71);
    p4_rec1=demod_bt_pkt1(1,64:117);
    p5_rec1=demod_bt_pkt1(1,118:length(bt_pkt_enc1));

    %###########################

    %p1_dec1=p1_rec1;
    p2_dec1=bchdeco(p2_rec1,k2,6);
    %p3_dec1=p3_rec1;
    %____________________________

     %p4 IS HEADER REPETION CODE
    %________________________________________
    for k=1:18
        counter=0;
    for l=1:3
        if p4_rec1(3*(k-1)+l)==1
            counter=counter+1;
        end
    end
    if counter >1
        p4_dec1(k)=1;
    else
        p4_dec1(k)=0;
    end
end
```

```
       %###########################
        p5_dec1=decode(p5_rec1',n1,k1,'hamming/binary');

         %*************DM3

    %p1_rec3=demod_bt_pkt3(1,1:4);
    p2_rec3=demod_bt_pkt3(1,1:63);          %IT FOR DH PKT ONE FROM DM
ONE FROM DH
    %p3_rec3=demod_bt_pkt3(1,68:71);
    p4_rec3=demod_bt_pkt3(1,64:117);
    p5_rec3=demod_bt_pkt3(1,118:length(bt_pkt_enc3));

    %###########################

    %p1_dec3=p1_rec3;
    p2_dec3=bchdeco(p2_rec3,k2,6);
    %p3_dec3=p3_rec3;
    %______________________________

     %p4 IS HEADER REPETION CODE
    %__________________________________________
    for k=1:18
        counter=0;
        for l=1:3
            if p4_rec3(3*(k-1)+l)==1
                counter=counter+1;
            end
        end
        if counter >1
            p4_dec3(k)=1;
        else
            p4_dec3(k)=0;
        end
    end

     %###########################
        p5_dec3=decode(p5_rec3',n1,k1,'hamming/binary');

        %*************DM5

    %p1_rec5=demod_bt_pkt5(1,1:4);
    p2_rec5=demod_bt_pkt5(1,1:63);          %IT FOR DH PKT ONE FROM DM
ONE FROM DH
    %p3_rec5=demod_bt_pkt5(1,68:71);
    p4_rec5=demod_bt_pkt5(1,64:117);
    p5_rec5=demod_bt_pkt5(1,118:length(bt_pkt_enc5));

    %###########################

    %p1_dec5=p1_rec5;
    p2_dec5=bchdeco(p2_rec5,k2,6);
    %p3_dec5=p3_rec5;
    %______________________________
```

```
 %p4 IS HEADER REPETION CODE
%_____________________________________________
for k=1:18
    counter=0;
    for l=1:3
        if p4_rec5(3*(k-1)+l)==1
            counter=counter+1;
        end
    end
    if counter >1
        p4_dec5(k)=1;
    else
        p4_dec5(k)=0;
    end
end

 %##########################
      p5_dec5=decode(p5_rec5',n1,k1,'hamming/binary');

%################################################################
%@################################################################
       bt_pkt_dec1=[p2_dec1,p4_dec1,p5_dec1'];
       bt_pkt_dec3=[p2_dec3,p4_dec3,p5_dec3'];
       bt_pkt_dec5=[p2_dec5,p4_dec5,p5_dec5'];
 %##################################################################
     x1=0;
     x2=0;
     x3=0;
       x1=sum(xor(bt_pkt_dec1, bt_pkt1));
       error_dm1=error_dm1+x1;
       x2=sum(xor(bt_pkt_dec3, bt_pkt3));
       error_dm3=error_dm3+x2;
       x3=sum(xor(bt_pkt_dec5, bt_pkt5));
       error_dm5=error_dm5+x3;
       if x1>0
           per1=per1+1;
       end
            if x2>0
           per2=per2+1;
       end
        if x3>0
           per3=per3+1;
       end

%##############################################################

%###############################################################

    end

%##############################################################

%##############################################################
```

```
        ber_dm1(j)=error_dm1/(length(bt_pkt1)*N)
        ber_dm3(j)=error_dm3/(length(bt_pkt3)*N) %NO OF ITERATION*NO
OF BITS IN PKT
        ber_dm5(j)=error_dm5/(length(bt_pkt5)*N) %NO OF ITERATION*NO
OF BITS IN PKT

        per_dm1(j)=per1/N
        per_dm3(j)=per2/N
        per_dm5(j)=per3/N

        per_Ndm1(j)=per1
        per_Ndm3(j)=per2
        per_Ndm5(j)=per3

        TH_per_dm1(j)=(136*(1-per_dm1(j)))/((1+1)*625)%
        TH_per_dm3(j)=(856*(1-per_dm3(j)))/((3+1)*625)%
        TH_per_dm5(j)=(1568*(1-per_dm5(j)))/((5+1)*625)%

    end
      figure(1);
       semilogy(snr,ber_dm1,'^-k');
       hold on
       semilogy(snr,ber_dm3,'O-r');
       hold on
       semilogy(snr,ber_dm5,'S-b');
       legend('DM1 Packets[PL=136]','DM3 Packets[PL=856]','DM5
Packets[PL=1568]');
       title('SNR  TO  BER CLASSIC 2DMx PACKETSF AWGN (Ham-7,4)');
       xlabel('Eb/No');
       ylabel('BER');
       grid on;
       figure(2);
       semilogy(snr,per_dm1,'*-k');
       hold on
       semilogy(snr,per_dm3,'+-r');
       hold on
       semilogy(snr,per_dm5);
       legend('DM1 Packets[PL=136]','DM3 Packets[PL=856]','2DM5
Packets[PL=1568]');
       title('SNR  TO  PER  CLASSIC DMx PACKETS AWGN (Ham-7,4)');
       xlabel('Eb/No');
       ylabel('PER');
       grid on
       figure(3);
       semilogy(snr,per_Ndm1,'*-k');
       hold on
       semilogy(snr,per_Ndm3,'+-r');
       hold on
       semilogy(snr,per_Ndm5);
       legend('DM1 Packets[PL=136]','DM3 Packets[PL=856]','DM5
Packets[PL=1568]');
       title('SNR  TO  NPL  CLASSIC DMx PACKETS AWGN (Ham-7,4)');
       xlabel('Eb/No');
       ylabel('No OF PACKETS LOSSES (NPL)');
       grid on
```

```
        figure(22);
       semilogy(per_dm1,TH_per_dm1,'s-k');
       hold on
       semilogy(per_dm3,TH_per_dm3,'d-k');
       hold on
       semilogy(per_dm5,TH_per_dm5,'^-r')
       legend('DM1 Packets[PL=136]','DM3 Packets[PL=856]','DM5
Packets[PL=1568]');
       title('PER  TO  Throughput  CLASSIC DMx PACKETS AWGN (Ham-
7,4)');
       xlabel('PER');
       ylabel('Throughput Mbps');
       grid on
       figure(33);
       semilogy(snr,TH_per_dm1,'s-k');
       hold on
       semilogy(snr,TH_per_dm3,'d-k');
       hold on
       semilogy(snr,TH_per_dm5,'^-r');
       legend('DM1 Packets[PL=136]','DM3 Packets[PL=856]','DM5
Packets[PL=1568]');
       title('SNR  TO  Throughput  CLASSIC DMx PACKETS AWGN (Ham-
7,4)');
        xlabel('SNR');
       ylabel('Throughput Mbps');
       grid on
```

#BCH (15, 7) code simulation Matlab code#

```
clc
clear all
close all
 %_-_-__________________BLUEOOTH SIMULATION__________________-_-_-_-%

 %%%%%%%%%%%%%%%%%BCH code (15, 7) SImulation%%%%%%%%%%%%%%%%%

    %_________________________________________________________
    %***********#########INPUT VARIABLE

    segm1=10 ;% input('the no of k data in of encoder =  ');
%segma1=10
    segm2=15;% input('the no of k data in of decoder =  ');
%segma2=15
    n2=63    ;% input('the no of codeword bch-access  =  ');
%n2=64
    k2=30    ;% input('the no of k data bch encoder =  ');          %
k2=30
    N=input('the NO of iteration                       =  ');
    n1=15;%input('the NO of codeword length  ham(15,11) n1=  ');
    k1=7;%input('the NO of dataword length  ham(15,11) k1=  ');
    t =2;%input('the no of corrected error bch(31,21)  t =  ');
```

```
snr=[0:1:10];
%snr=[0:5:35];
y=length(snr);
snrabs=[];
counter_iter=0;
for j=1:y


       counter_iter=counter_iter+1
       snrabs(j)=10^(snr(j)/10);
       sigma1=1/sqrt(2*snrabs(j));%    k_1=224   &  n_1=373
DM_PACKETS
       sigma3=1/sqrt(2*snrabs(j));  %k_3=1064   &  n_1=1613
DM_PACKETS
       sigma5=1/sqrt(2*snrabs(j));  % k_5=1848   &  n_1=2853
DM_PACKETS
       error_dh1=0;
       error_dm1=0;
       error_dh3=0;
       error_dm3=0;
       error_dh5=0;
       error_dm5=0;
       per1=0;
       per2=0;
       per3=0;

       for i=1:N


    %p1=round(rand(1,4));                  %preample
    p2=round(rand(1,30));                 %sync word LAP=24+6
    %p3=round(rand(1,4));                  %trailer
    p4=round(rand(1,18));                 %header
    dm1=round(rand(1,112));               %payload dm1 packet fec
(15,10)ham
    dm3=round(rand(1,700));              %payload dm3 packet ,,,,,,,,,,
    dm5=round(rand(1,1281));              %payload dm5 packet,,,,,,,,,,,
    %----------------------------------------
    %dh1=round(rand(1,240));              %payload dh1 packet  no fec
    %dh3=round(rand(1,1500));              %payload dh3 packet ,,,,,
    %dh5=round(rand(1,2745));              %payload dh5 packet  ,,,,,
    %##############################################################
    bt_pkt1=[p2,p4,dm1];
    bt_pkt3=[p2,p4,dm3];
    bt_pkt5=[p2,p4,dm5];

    %---------------------------------=



    %bt_pkth1=[p1,p2,p3,p4,dh1];
    %bt_pkth3=[p1,p2,p3,p4,dh3];
    %bt_pkth5=[p1,p2,p3,p4,dh5];
    %************************************************************
    %---------------------------------
```

```
% p2 sync word use code bch code(64,30)

%p2=gf(p2);             %BCH CODE ACCESS CODE PART IN GENERAL FORM
%[genpoly,t]=bchgenpoly(n2,k2);
 %pg = bchpoly(n2, k2);
 p2_encode=bchenco(p2,n2,k2);

%*****************************************
% P4 IS HEADER USE REPETITION CODE 1/3
for k=1:18

    for l=1:3
        p4_encode(3*(k-1)+l)=p4(k);
    end
end

%*************************************************
%P5=PDM OR PDH OR OTHER DM USE HAMM.CODE (15,10)
% BUT DH NO FEC
%dm1_encode=encode(dm1,n1,k1,'hamming/binary');
%ENCODING(X,G,SEGM) IT IS FUNCTION
dm=dm1;
dm_enc=encoding1(dm,n1,k1);
dm1_encode=dm_enc;
%dm3_encode=encode(dm3,n1,k1,'hamming/binary');
 dm=dm3;
dm_enc=encoding1(dm,n1,k1);
dm3_encode=dm_enc;
%dm5_encode=encode(dm5,n1,k1,'hamming/binary');
dm=dm5;
dm_enc=encoding1(dm,n1,k1);
dm5_encode=dm_enc;

%*****************************************************
%so the bt_pkt after encoding & before sending  it as will be
%shown

bt_pkt_enc1=[p2_encode,p4_encode,dm1_encode];
bt_pkt_enc3=[p2_encode,p4_encode,dm3_encode];
bt_pkt_enc5=[p2_encode,p4_encode,dm5_encode];
%-------------------------------------
%bt_pkt_ench1=[p1,p2_encode,p3,p4_encode,dh1];
%bt_pkt_ench3=[p1,p2_encode,p3,p4_encode,dh3];
%bt_pkt_ench5=[p1,p2_encode,p3,p4_encode,dh5];


%******************MOD. AS BPSK FOR ALL PACKET TYPE*****
%***************** MODULATION
mod_bt_pkt1=2*bt_pkt_enc1-1;
mod_bt_pkt3=2*bt_pkt_enc3-1;
mod_bt_pkt5=2*bt_pkt_enc5-1;

%-----------------------------

%mod_bt_pkth1=2*bt_pkt_ench1-1;
%mod_bt_pkth3=2*bt_pkt_ench3-1;
%mod_bt_pkth5=2*bt_pkt_ench5-1;
```

```
%###################################################################
%###################################################################
%###################################################################
    %########################  1- AWGN Channel
############################

    rec_bt_pkt1=mod_bt_pkt1+sigma1*randn(1,length(bt_pkt_enc1));
    rec_bt_pkt3=mod_bt_pkt3+sigma1*randn(1,length(bt_pkt_enc3));
    rec_bt_pkt5=mod_bt_pkt5+sigma1*randn(1,length(bt_pkt_enc5));

%###################################################################
    %**##################**2-SLOW FADING  ##################

    %deep_fad=sqrt(0.5*(randn(1,1)).^2+0.5*(randn(1,1)).^2);
   % ##############
   % bt_pkt_d_fad1=mod_bt_pkt1*deep_fad;
    %rec_bt_pkt1=bt_pkt_d_fad1+sigma1*randn(1,length(bt_pkt_enc1));
    %##################
    %bt_pkt_d_fad3=mod_bt_pkt3*deep_fad;
    %rec_bt_pkt3=bt_pkt_d_fad3+sigma3*randn(1,length(bt_pkt_enc3));
    %###################
    %bt_pkt_d_fad5=mod_bt_pkt5*deep_fad;
    %rec_bt_pkt5=bt_pkt_d_fad5+sigma5*randn(1,length(bt_pkt_enc5));


%###################################################################
    %############### 3- FLAT FAST BLOCK FADING CHANEL
###############
                              % STANDARD ENCODED PACKETS
   %
prf_fad1=sqrt((0.5*(randn(1,length(bt_pkt_enc1))).^2)+(0.5*(randn(1,le
ngth(bt_pkt_enc1))).^2));
   % bt_pkt_p_fad1=mod_bt_pkt1.*prf_fad1;
   % rec_bt_pkt1=bt_pkt_p_fad1+sigma1*randn(1,length(bt_pkt_enc1));


%prf_fad3=sqrt((0.5*(randn(1,length(bt_pkt_enc3))).^2)+(0.5*(randn(1,l
ength(bt_pkt_enc3))).^2));
    %bt_pkt_p_fad3=mod_bt_pkt3.*prf_fad3;
    %rec_bt_pkt3=bt_pkt_p_fad3+sigma3*randn(1,length(bt_pkt_enc3));

   %
prf_fad5=sqrt((0.5*(randn(1,length(bt_pkt_enc5))).^2)+(0.5*(randn(1,le
ngth(bt_pkt_enc5))).^2));
   % bt_pkt_p_fad5=mod_bt_pkt5.*prf_fad5;
   % rec_bt_pkt5=bt_pkt_p_fad5+sigma5*randn(1,length(bt_pkt_enc5));

%###################################################################

%###################################################################
    %###############   4- CORRELATED FADING     ###############
                              % STANDARD ENCODED PACKETS
```

```
% [row,column]=size(bt_pkt_enc1);
%a=jack_fading(column);
%rec_bt_pkt1=a.*(mod_bt_pkt1)+sigma1*randn(1,length(bt_pkt_enc1));

%[row,column]=size(bt_pkt_enc3);
%a=jack_fading(column);
%rec_bt_pkt3=a.*(mod_bt_pkt3)+sigma3*randn(1,length(bt_pkt_enc3));

%[row,column]=size(bt_pkt_enc5);
%a=jack_fading(column);
%rec_bt_pkt5=a.*(mod_bt_pkt5)+sigma5*randn(1,length(bt_pkt_enc5));
%####################################################################
%**********************DEMODULATION

% demod_bt_pkt1=rec_bt_pkt1>0;
%________________
for r=1:length(rec_bt_pkt1)
    if rec_bt_pkt1(r)>0
      demod_bt_pkt1(r)=1;
else
      demod_bt_pkt1(r)=0;
end
end
    %demod_bt_pkt3=rec_bt_pkt3>0;
    %_______________________
    for r=1:length(rec_bt_pkt3)
        if rec_bt_pkt3(r)>0

    demod_bt_pkt3(r)=1;
else
    demod_bt_pkt3(r)=0;
end
end
   % demod_bt_pkt5=rec_bt_pkt5>0;
   %___________________________
    for r=1:length(rec_bt_pkt5)
        if rec_bt_pkt5(r)>0

    demod_bt_pkt5(r)=1;
else
    demod_bt_pkt5(r)=0;
end
end
   %demod_bt_pkth3=rec_bt_pkth3>0;
   %demod_bt_pkth5=rec_bt_pkth5>0;

   %**********************DECODING EACH PART ALONE
   %----------------------------
   %p1_rech1=demod_bt_pkth1(1,1:4);
   %p2_rech1=demod_bt_pkth1(1,5:67);    %IT FOR DH PKT ONE FROM DM
ONE FROM DH
   %p3_rech1=demod_bt_pkth1(1,68:71);
   %p4_rech1=demod_bt_pkth1(1,72:125);
   %p5_rech1=demod_bt_pkth1(1,126:365);
   %________________________________
   %DECODING FOR DMI OMLY HENCE
   % P1 PREAMPLE NO FEC
```

```
 %p1_dec1=p1_rec1;
% NOW DECODE THE PART OF BCH CODE WHICH FOR ACCESS CODE
%p2_dec1=bchdeco(p2_rec1,k2,6);
%p3   TRAILER NO FEC
%p3_dec1=p3_rec1;
%p4 IS HEADER REPETION CODE
%_______________________________________
%for k=1:18
 %   counter=0;
  %  for l=1:3
   %      if p4_rec1(3*(k-1)+l)==1
    %          counter=counter+1;
    %    end
    %end
    %if counter >1
     %   p4_dec1(k)=1;
     %else
     %   p4_dec1(k)=0;
     %end
     %end
%__________________________________________________

%NOW HAMM CODE FOR DM PKT BUT DIRECT FOR DH PKT

%p5_dec1=decoding(p5_rec1,H,segm2);    % IT IS FUNCTION  H IS
PARITY- CHECK MATRIX

% so the pkt after THAT IS
%bt_pkt_dec1=[p1_dec1,p2_dec1,p3_dec1,p4_dec1,p5_dec1];
%<============================

%************DM1

%p1_rec1=demod_bt_pkt1(1,1:4);
p2_rec1=demod_bt_pkt1(1,1:63);       %IT FOR DH PKT ONE FROM DM
ONE FROM DH
%p3_rec1=demod_bt_pkt1(1,68:71);
p4_rec1=demod_bt_pkt1(1,64:117);
p5_rec1=demod_bt_pkt1(1,118:length(bt_pkt_enc1));

%##########################

%p1_dec1=p1_rec1;
p2_dec1=bchdeco(p2_rec1,k2,6);
%p3_dec1=p3_rec1;
%____________________________

 %p4 IS HEADER REPETION CODE
%_______________________________________
for k=1:18
    counter=0;
    for l=1:3
        if p4_rec1(3*(k-1)+l)==1
            counter=counter+1;
        end
    end
```

```
        if counter >1
            p4_dec1(k)=1;
        else
            p4_dec1(k)=0;
        end
    end

         %###########################
         % p5_dec1=decode(p5_rec1',n1,k1,'hamming/binary');
         dm_rec=p5_rec1;
         dm_dec=decoding1(dm_rec,n1,k1,t);
         p5_dec1=dm_dec;
           %*************DM3

    %p1_rec3=demod_bt_pkt3(1,1:4);
    p2_rec3=demod_bt_pkt3(1,1:63);       %IT FOR DH PKT ONE FROM DM
ONE FROM DH
    %p3_rec3=demod_bt_pkt3(1,68:71);
    p4_rec3=demod_bt_pkt3(1,64:117);
    p5_rec3=demod_bt_pkt3(1,118:length(bt_pkt_enc3));

    %###########################

    %p1_dec3=p1_rec3;
    p2_dec3=bchdeco(p2_rec3,k2,6);
    %p3_dec3=p3_rec3;
    %____________________________

     %p4 IS HEADER REPETION CODE
    %_______________________________________
    for k=1:18
        counter=0;
        for l=1:3
            if p4_rec3(3*(k-1)+l)==1
                counter=counter+1;
            end
        end
        if counter >1
            p4_dec3(k)=1;
        else
            p4_dec3(k)=0;
        end
    end

    %###########################
         %p5_dec3=decode(p5_rec3',n1,k1,'hamming/binary');
         dm_rec=p5_rec3;
        dm_dec=decoding1(dm_rec,n1,k1,t);
        p5_dec3=dm_dec;
         %*************DM5

    %p1_rec5=demod_bt_pkt5(1,1:4);
    p2_rec5=demod_bt_pkt5(1,1:63);       %IT FOR DH PKT ONE FROM DM
ONE FROM DH
    %p3_rec5=demod_bt_pkt5(1,68:71);
    p4_rec5=demod_bt_pkt5(1,64:117);
    p5_rec5=demod_bt_pkt5(1,118:length(bt_pkt_enc5));
```

```
         %##########################

         %p1_dec5=p1_rec5;
         p2_dec5=bchdeco(p2_rec5,k2,6);
         %p3_dec5=p3_rec5;
         %____________________________
          %p4 IS HEADER REPETION CODE
         %________________________________________
         for k=1:18
             counter=0;
             for l=1:3
                 if p4_rec5(3*(k-1)+l)==1
                     counter=counter+1;
                 end
             end
             if counter >1
                 p4_dec5(k)=1;
             else
                 p4_dec5(k)=0;
             end
         end

          %##########################

         % p5_dec5=decode(p5_rec5',n1,k1,'hamming/binary');
          dm_rec=p5_rec5;
         dm_dec=decoding1(dm_rec,n1,k1,t);
         p5_dec5=dm_dec;
 %###############################################################
 %@################################################################
         bt_pkt_dec1=[p2_dec1,p4_dec1,p5_dec1];
         bt_pkt_dec3=[p2_dec3,p4_dec3,p5_dec3];
         bt_pkt_dec5=[p2_dec5,p4_dec5,p5_dec5];
  %###################################################################
       x1=0;
       x2=0;
       x3=0;
         x1=sum(xor(bt_pkt_dec1, bt_pkt1));
         error_dm1=error_dm1+x1;
         x2=sum(xor(bt_pkt_dec3, bt_pkt3));
         error_dm3=error_dm3+x2;
         x3=sum(xor(bt_pkt_dec5, bt_pkt5));
         error_dm5=error_dm5+x3;
         if x1>0
             per1=per1+1;
         end
              if x2>0
             per2=per2+1;
         end
          if x3>0
             per3=per3+1;
         end

%##############################################################

%###############################################################
```

```
    end
%##############################################################

%##############################################################
        ber_dm1(j)=error_dm1/(length(bt_pkt1)*N)
        ber_dm3(j)=error_dm3/(length(bt_pkt3)*N) %NO OF ITERATION*NO
OF BITS IN PKT
        ber_dm5(j)=error_dm5/(length(bt_pkt5)*N) %NO OF ITERATION*NO
OF BITS IN PKT

        per_dm1(j)=per1/N
        per_dm3(j)=per2/N
        per_dm5(j)=per3/N

        per_Ndm1(j)=per1
        per_Ndm3(j)=per2
        per_Ndm5(j)=per3

        TH_per_dm1(j)=(112*(1-per_dm1(j)))/((1+1)*625)%
        TH_per_dm3(j)=(700*(1-per_dm3(j)))/((3+1)*625)%
        TH_per_dm5(j)=(1281*(1-per_dm5(j)))/((5+1)*625)%
    end
      figure(1);
       semilogy(snr,ber_dm1,'^-k');
       hold on
       semilogy(snr,ber_dm3,'O-r');
       hold on
       semilogy(snr,ber_dm5,'S-b');
       legend('DM1 Packets[PL=112]','DM3 Packets[PL=700]','DM5
Packets[PL=1281]');
       title('SNR  TO  BER CLASSIC DM PACKETSF AWGN (BCH-15,7');
       xlabel('Eb/No');
       ylabel('BER');
       grid on;
       figure(2);
       semilogy(snr,per_dm1,'*-k');
       hold on
       semilogy(snr,per_dm3,'+-r');
       hold on
       semilogy(snr,per_dm5);
       legend('DM1 Packets[PL=112]','DM3 Packets[PL=700]','DM5
Packets[PL=1281]');
       title('SNR  TO  PER  CLASSIC DM PACKETS AWGN bch(15,7)');
       xlabel('Eb/No');
       ylabel('PER');
       grid on
       figure(3);
       semilogy(snr,per_Ndm1,'*-k');
       hold on
       semilogy(snr,per_Ndm3,'+-r');
       hold on
       semilogy(snr,per_Ndm5);
       legend('DM1 Packets[PL=112]','DM3 Packets[PL=700]','DM5
Packets[PL=1281]');
```

```
        title('SNR  TO  NPL  CLASSIC DM PACKETS AWGN bch(15,7)');
        xlabel('Eb/No');
        ylabel('No OF PACKETS LOSSES');
        grid on

         figure(22);
        semilogy(per_dm1,TH_per_dm1,'s-k');
        hold on
        semilogy(per_dm3,TH_per_dm3,'d-k');
        hold on
        semilogy(per_dm5,TH_per_dm5,'^-r')
        legend('DM1 Packets[PL=112]','DM3 Packets[PL=700]','DM5
Packets[PL=1281]');
        title('PER  TO  Throughput  CLASSIC DMx PACKETS AWGN
bch(15,7)');
        xlabel('PER');
        ylabel('Throughput Mbps');
        grid on
        figure(33);
        semilogy(snr,TH_per_dm1,'s-k');
        hold on
        semilogy(snr,TH_per_dm3,'d-k');
        hold on
        semilogy(snr,TH_per_dm5,'^-r');
        legend('DM1 Packets[PL=112]','DM3 Packets[PL=700]','DM5
Packets[PL=1281]');
        title('SNR  TO  Throughput  CLASSIC DMx PACKETS AWGN
bch(15,7)');
         xlabel('SNR');
        ylabel('Throughput Mbps');
        grid on
```

### *4.5.3 Performance Comparison*

This section presents a quick comparison of the $DM_x$ packet performance in standard case and proposed error control code scheme cases.

Then, several simulation experiments are carried out to investigate the effects of using proposed coding schemes on the $DM_x$ performance compared with $DH_x$ performance. Figures 4.14, 4.15, and 4.16 show simulation results of using Hamming (7, 4), cyclic (15, 11), and BCH (15, 7) coding schemes in PL of $DM_x$ packets over AWGN channel. These figures compare the performance of $DM_1$, $DM_3$, and $DM_5$ packets in case of different coding schemes and uncoded packets $DH_1$, $DH_3$, and $DH_5$, respectively.

Figure 4.14 shows the comparison of the performance of uncoded $DH_1$ packets with the performance of encoded $DM_1$ packets [standard case, cyclic code, Hamming code (7, 4), and BCH code (15, 7)] over AWGN channel. As shown in this figure, by using these schemes of error control codes, $DM_1$ performs better than $DH_1$ and $DM_1$ with shortened Hamming code. At PER $10^{-3}$, the values of SNR are

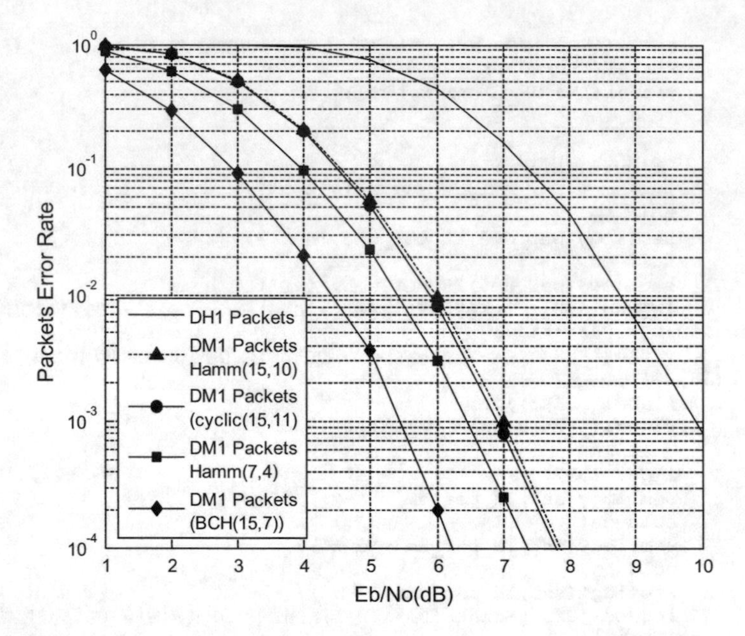


**Fig. 4.14** Simulated PER of $DM_1$ and $DH_1$ on AWGN channel

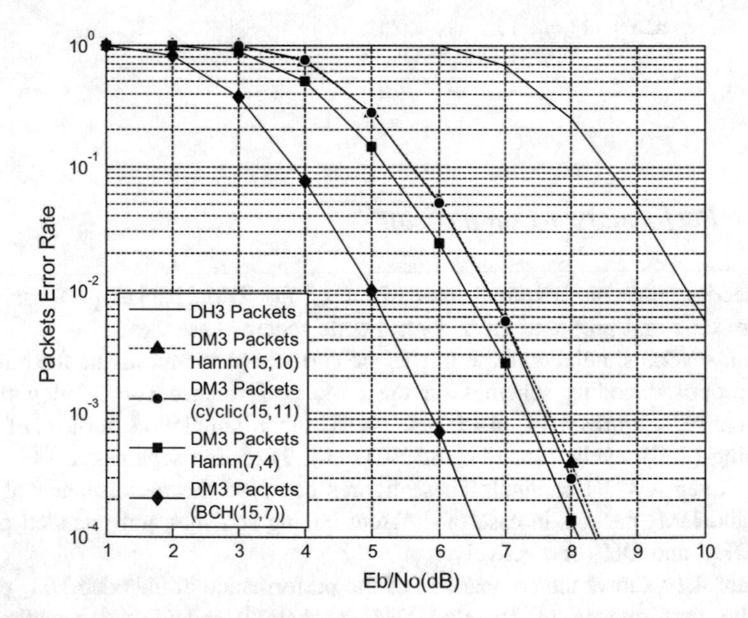


**Fig. 4.15** Simulated PER of $DM_3$ performance and $DH_3$ over AWGN channel

5.5, 6.4, 6.9, 7, and 9.99 dB for BCH, Hamming (7, 4), cyclic (15, 11), shortened Hamming (15, 10) codes of $DM_1$ packets and uncoded $DH_1$ packets, respectively, as provided in Table 4.1.

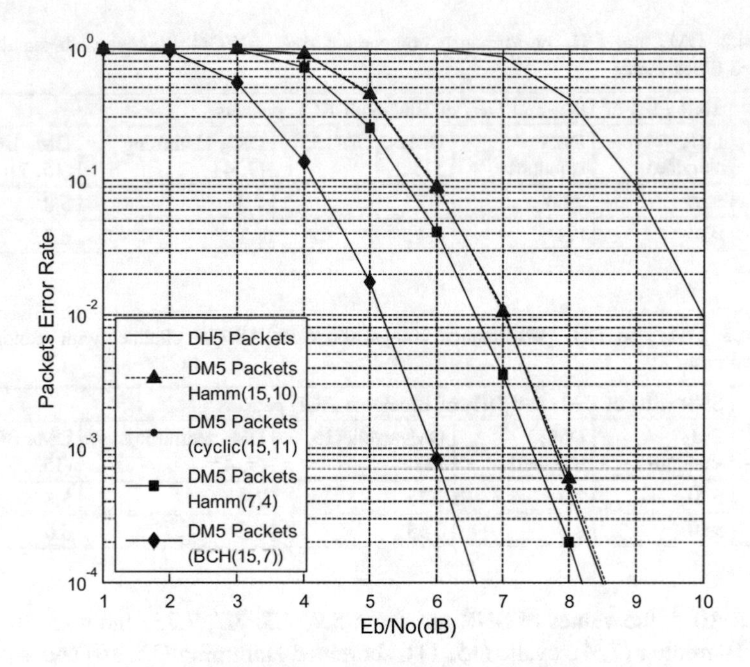


**Fig. 4.16** Simulated PER of $DM_5$ performance and $DH_5$ on AWGN channel

**Table 4.1** $DM_1$ and $DH_1$ performance comparison over AWGN channel with standard and proposed block codes

| PER | SNR (dB) of $DM_1$ and $DH_1$ of Bluetooth ACL packets | | | | |
|---|---|---|---|---|---|
| | $DH_1$ standard | $DM_1$ standard | $DM_1$ cyclic(15, 11) | $DM_1$ Hamming (7, 4) | $DM_1$ BCH (15, 7) |
| $10^{-3}$ | 9.99 | 6.9 | 7 | 6.4 | 5.5 |
| $10^{-4}$ | >10 | 7.8 | 7.9 | 7.3 | 6.2 |

Figure 4.15 shows the comparison between the performance of $DM_3$ packets and $DH_3$ packets that is over AWGN channel. $DM_3$ packets perform better than $DH_3$. The best performance of $DM_3$ is given by using BCH code (15,7). As shown in Fig. 4.15 at PER $10^{-3}$, the values of SNR are 5.8, 7.2, 7.7, 7.66, and over 10 dB for BCH, Hamming (7, 4), cyclic (15, 11), shortened Hamming (15, 10) codes of $DM_3$ packets and uncoded $DH_3$ packets, respectively. As shown, performance of $DM_1$ and $DH_1$ is better than $DM_3$ and $DH_3$, respectively, as shown in Table 4.2.

As shown in Fig. 4.16, it gives comparison between the performance of $DM_5$ packets and $DH_5$ packets that is over AWGN channel. $DM_5$ packets perform better than $DH_5$. The best performance of $DM_3$ is given by using BCH code (15, 7).

**Table 4.2** $DM_3$ and $DH_3$ performance comparison over AWGN channel with standard and proposed block codes

| PER | SNR (dB) of $DM_3$ and $DH_3$ of Bluetooth ACL packets | | | | |
|---|---|---|---|---|---|
| | $DH_3$ standard | $DM_3$ standard | $DM_3$ cyclic(15, 11) | $DM_3$ Hamming (7, 4) | $DM_3$ BCH (15, 7) |
| $10^{-3}$ | >10 | 7.66 | 7.7 | 7.2 | 5.8 |
| $10^{-4}$ | >10 | 8.3 | 8.4 | 8.1 | 6.5 |

**Table 4.3** $DM_5$ and $DH_5$ performance comparison over AWGN channel with standard and proposed case

| PER | SNR (dB) of $DM_5$ and $DH_5$ of Bluetooth ACL packets | | | | |
|---|---|---|---|---|---|
| | $DH_5$ standard | $DM_5$ standard | $DM_5$ cyclic(15, 11) | $DM_5$ Hamming (7, 4) | $DM_5$ BCH (15, 7) |
| $10^{-3}$ | >10 | 7.7 | 7.75 | 7.4 | 5.8 |
| $10^{-4}$ | >10 | 8.5 | 8.55 | 8.2 | 6.6 |

At PER $10^{-3}$, the values of SNR are about 5.9, 7.3, 7.7, 7.75, and over 10 dB for BCH, Hamming (7, 4), cyclic (15, 11), shortened Hamming (15, 10) codes of $DM_5$ packets and uncoded $DH_5$ packets, respectively. These figures reveal that the performance of $DM_5$ and $DM_3$ are very close, as shown in Table 4.3.

These figures reveal that the cyclic (15, 11) and Hamming (15, 10) codes have very closed results over AWGN channel. Hamming code (7, 4) has a better performance than Hamming code (15, 10). If redundancy length is tolerated, the best performance of $DM_x$ packets is given by using BCH code (15, 7) where performance gain of $DM_x$ packets over $DH_x$ packets is more than 4.5 dB, as shown in Table 4.3.

The following figures show the performance comparison between the performance of $DM_x$ packets (both standard case and other block code types) and $DH_x$ packets over a Rayleigh-flat fading channel.

Figure 4.17 shows the comparison of the performance of uncoded $DH_1$ packets with the performance of encoded $DM_1$ packets (both standard case and other block code types) that is over Rayleigh-flat fading channel. As shown in this figure, by using these schemes of error control code schemes, $DM_1$ performs better than $DH_1$ and $DM_1$ with shortened Hamming code. At PER $10^{-3}$, the values of SNR are 30.9, 32.3, 33.2, 33.2, and over 35 dB for BCH, Hamming (7, 4), cyclic (15, 11), shortened Hamming (15, 10) codes of $DM_1$ packets and uncoded $DH_1$ packets, respectively [18], as shown in Table 4.4.

Figure 4.18 shows the comparison of the performance of uncoded $DH_3$ packets with the performance of encoded $DM_3$ packets (both standard case and other block code types) over Rayleigh-flat fading channel. As shown in this figure, by using

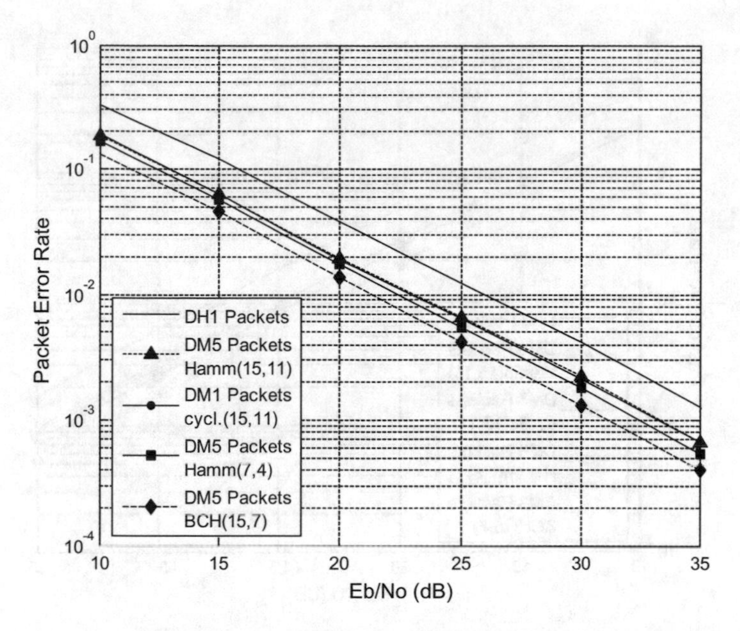


**Fig. 4.17** Simulated PER of $DM_1$ performance and $DH_1$ on Rayleigh-flat fading channel

**Table 4.4** $DM_1$ and $DH_1$ performance comparison over Rayleigh-flat fading channel with standard and proposed block codes

| PER | SNR (dB) of $DM_1$ and $DH_1$ of Bluetooth ACL packets | | | | |
|---|---|---|---|---|---|
| | $DH_1$ standard | $DM_1$ standard | $DM_1$ cyclic(15, 11) | $DM_1$ Hamming (7, 4) | $DM_1$ BCH (15, 7) |
| $10^{-2}$ | 26.1 | 23.23 | 23.1 | 22.5 | 21.5 |
| $10^{-3}$ | >35 | 33.2 | 33.2 | 32.3 | 30.9 |

these schemes of error control code schemes, $DM_3$ performs better than $DH_3$ and $DM_3$ with shortened Hamming code. At PER $10^{-3}$, the values of SNR are 32.3, 34, 34.8, 34.8, and over 35 dB for BCH, Hamming (7, 4), cyclic (15, 11), shortened Hamming (15, 10) codes of $DM_3$ packets and uncoded $DH_3$ packets, respectively, as shown in Table 4.5.

In the same manner in case of $DM_5$ (the same coding schemes) and $DH_5$, the result of performance comparison is shown in Fig. 4.19 and these figures reveal that the performance of $DM_5$ and $DM_3$ is very close.

At PER $10^{-3}$, the values of SNR are 32.5, 34.5, 35.1, 35, and over 35 dB for BCH, Hamming (7, 4), cyclic (15, 11), shortened Hamming (15, 10) codes of $DM_5$ packets and uncoded $DH_5$ packets, respectively, as shown in Table 4.6.

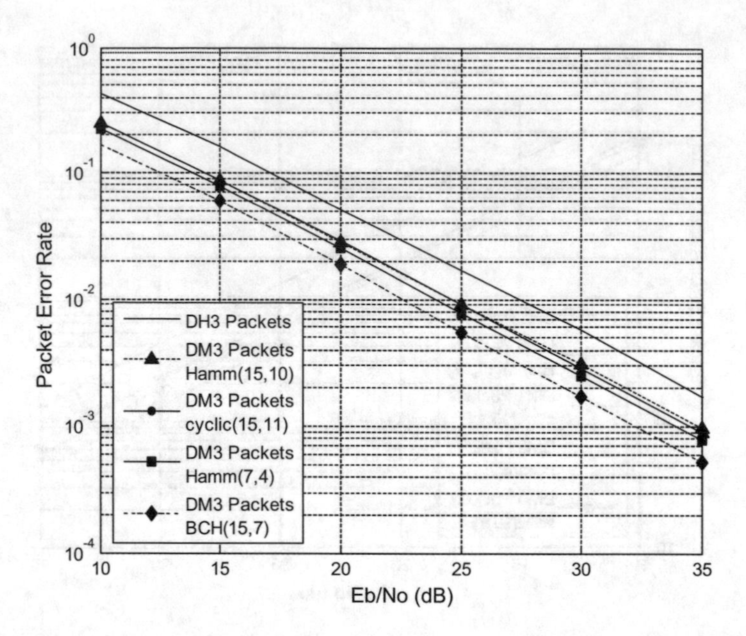


**Fig. 4.18** Simulated PER of $DM_3$ performance and $DH_3$ on Rayleigh-flat fading channel

**Table 4.5** $DM_3$ and $DH_3$ performance comparison over Rayleigh-flat fading channel with standard and proposed block codes

| PER | SNR (dB) of $DM_3$ and $DH_3$ of Bluetooth ACL packets | | | | |
|---|---|---|---|---|---|
| | $DH_3$ standard | $DM_3$ standard | $DM_3$ cyclic(15, 11) | $DM_3$ Hamming (7, 4) | $DM_3$ BCH (15, 7) |
| $10^{-2}$ | 27.6 | 24.8 | 24.81 | 24.1 | 22.5 |
| $10^{-3}$ | >35 | 34.8 | 34.81 | 34 | 32.3 |

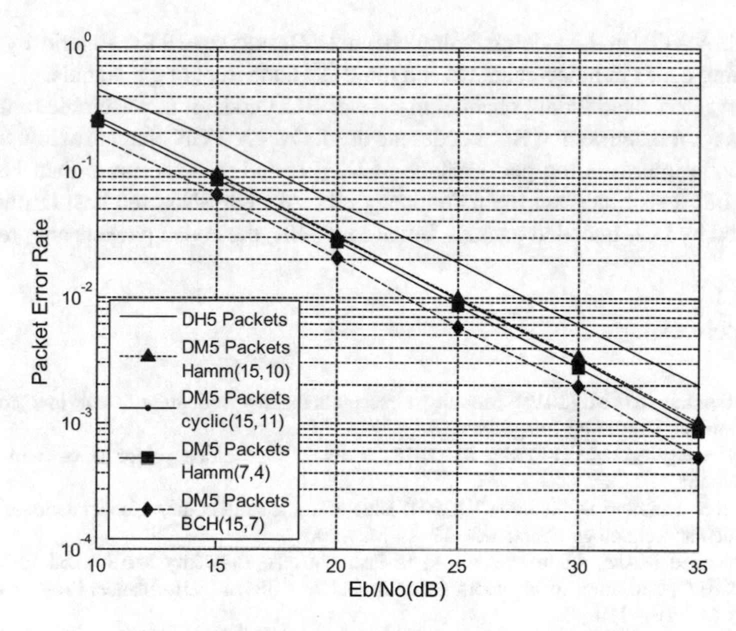


**Fig. 4.19** Simulated PER of $DM_5$ performance and $DH_5$ on Rayleigh-flat fading channel

**Table 4.6** $DM_5$ and $DH_5$ performance comparison over Rayleigh-flat fading channel with standard and proposed block codes

| PER | SNR (dB) of $DM_5$ and $DH_5$ of Bluetooth ACL packets | | | | |
|---|---|---|---|---|---|
| | $DH_5$ standard | $DM_5$ standard | $DM_5$ cyclic(15, 11) | $DM_5$ Hamming (7, 4) | $DM_5$ BCH (15, 7) |
| $10^{-2}$ | 27.6 | 25 | 25 | 24.3 | 22.8 |
| $10^{-3}$ | >35 | 35 | 35.02 | 34.5 | 32.5 |

# 4.6 Summary

These previous simulation results reveal that the BCH code gives a good difference in performance for $DM_x$ packets in the case of AWGN channel as compared to $DH_x$ packet performance over the same channel. Results reveal that the use of cyclic code gives performance the same Hamming code (15, 10) with lower redundancy length. Shorter Hamming code performs better than longer ones. Performance of shorter packets ($DM_1$) is better than longer packets ($DM_3$, $DM_5$).

Our simulation investigates the performance of Bluetooth system with the existing Bluetooth error control codes. We concerned with studying ACL Bluetooth packets, which are encoded packets ($DM_x$) and uncoded packets ($DH_x$). The study of Bluetooth ACL packets over AWGN and fading channels is presented. This chapter presents some possible frame formats for Bluetooth depending on the distribution of the payload field between data and checksum. Hamming and BCH codes are investigated for both AWGN and Rayleigh fading channels. Results reveal the superiority of short Hamming codes in noisy channels and the BCH code in fading channels.

Over interleaved fading channel in case of BCH code, it is preferable to use $DM_5$ packets for transmission if BCH codes are used. Over AWGN channel at low SNR, the BCH code improves the performance of $DM_x$ coded packets, more than Hamming codes, but it will increase the redundancy bits. At high SNR, the best throughput is achieved by $DH_x$ uncoded packets, but at low SNR, the coded packets are preferable.

## References

1. El-Bendary MAM (2013) Mobility effects combating through efficient low complexity technique. Ciit- Digit Image Process
2. El-Bendary MAM, El-Tokhy M (2013) WSN ZigBee based performance. Ciit- J Wirel Commun
3. Galli S, Famolari D, Kodama T (2004) Bluetooth: channel coding Considerations. In: IEEE vehicular technology conference, 17–19 May 2004
4. Mohamed MAM, Abou El-Azm A, El-Fishawy NA, El-Tokhy MAR, Abd El-Samie FE (2008) Optimization of bluetooth frame format for efficient performance. Prog Electromagn Res M 1:101–110
5. El-Bendary MAM, Abou El-Azm AE, El-Fishawy NA, El-Tokhy MAR, Abd El-Samie FE (2013) Image transmission over mobile Bluetooth networks with enhanced data rate packets and chaotic interleaving. Wirel Netw
6. Jacobson M, Wetzel S Security weaknesses in bluetooth, online report, http://www.bell-labs.com/user/markusj/bluetooth.pdf
7. Conti A, Dardari D, Paolini G, Andrisano O (2003) Bluetooth and IEE 802.11b coexistence: analytical performance evaluation in fading channels. IEEE J Sel Areas Commun 21(2)
8. Hamdi KA (2002) Exact Probability of error BPSK communication links subjected to asynchronous interference in Rayleigh Fading Environment 50(10)
9. Proaskis JG (1989) Digital communication, 2nd edn. Mc Grow Hill
10. Golmie N, Van Dck RE, Soltanian A (2001) Interference of bluetooth and IEEE 802.11: simulation modeling and performance evaluation. In: Proceedings ACM international workshop on modeling, analysis, and simulation of wireless and mobile systems, Rome, Italy, July 2001
11. Haartsen JC, Zurbes S (1999) Bluetooth voice and data performance in 802.11 DS WLAN environments. Ericsson report, May 1999
12. Johansson P, Kapoor R, Kazantzidis M, Gerla M (2001) Bluetooth: an enabler for personal area networking. IEEE Netw Mag
13. Elliott EO (1963) Estimates of error rates for codes on burst-error channels. Bell Syst Tech J 42
14. IEEE 802.15.4 wpan-lr task group http://www.ieee802.org/15/pub/TG4.html
15. Golmie N, Chevrollier N, Elbakkouri I (2001) Interference aware bluetooth packet scheduling. In: GLOBECOM
16. Kasban H, El-Tokhy MAR, El-Bendary MAM (2014) Interleaved reed-solomon codes with code rate switching over wireless. Int J Inf Technol Comput Sci 16(1)
17. Mohamed MAM, El-Azm AEA, El-Fishawy NA, Shawky F Bluetooth performance improvement using convolutional codes. J Autom Syst Eng 3(1):1–10
18. El-Bendary MAM, El-Azm AEA, El-Fishawy NA, Shawky F, Abdelsamie F (2009) Throughput improvement over Bluetooth system through adaptive packets. IEEE Int Conf Comput Eng Syst, pp. 616–621
19. El-Bendary MAM, El-Azm AEA, El-Fishawy NA, Shawky F, El-Samie FE (2011) Embedded Throughput Improving of Low-rate EDR Packets for Lower-latency. World Acad Sci, Eng Technol, Int J Electr, Comput, Energ, Electron Commun Eng, 5(3):449–458

# Chapter 5
# WPAN Simulation Scenarios-2 with the Different Coding

## 5.1 Introduction

Mostly, the efficient performance requires complex error control schemes. The convolutional codes are complex error control technique compared with the block codes. In this chapter, the powerful error control scheme is employed for encoding the transmitted Bluetooth packets. The performance will enhance with overhead complexity. The proposed scenario is effective over the bad condition communications channels.

There are two types channel codes, block codes and convolutional codes. Block code is presented in previous chapters. The structure of convolutional codes encoders is different from block codes. In this chapter, we present convolutional code in Bluetooth frame structure modification and analyze the effects of this modification of the performance of Bluetooth system over AWGN channel and Rayleigh fading channel [1].

## 5.2 Convolutional Codes

In practical communication systems, Convolutional codes are widely utilized for combating the communications channel errors and improving the wireless communications system error performance. Convolutional code is unlike block code, where convolutional code converts the entire data stream (input of encoder) into one single code-work (output of encoder). In this case, the output of encoder bits *n* (codeword) not only depends on the current input of encoder bits but also depends on past input bits, where convolutional encoder operates on the input bits of information sequence continuously in a serial manner. There are many methods for decoding but the main decoding method for convolutional codes is based on the Viterbi algorithm. Before analyzing the results of new Bluetooth format performance over different channels, we will describe encoder of convolutional code structure briefly in [2].



M. A. M. El-Bendary, *Wireless Personal Communications*,
Signals and Communication Technology,
https://doi.org/10.1007/978-981-10-7131-7_5

The encoder of convolutional code with rate $R = 1/n$ consists of an M-stage shift register connected to modulo-2 adders, and the output is taken through multiplexer, which serializes the output of the adders. Let us assume that, $L$ bits sequence is the input of the convolutional encoder. So the output of this encoder will be $n\,(L + M)$ bits sequence, and then actual form of code rate of convolutional code is given by following equation.

$$R = L/n(L+M) \tag{5.1}$$

Before approximation of this Eq. (5.1), we will define its different terms as follows:

$R$ is code rate of convolutional code, which is ratio of input of encoder to its output.
$L$ is the number of message bits sequence, which is the input of encoder length.
$n$ is the number of parallel output encoded bits at one time interval.
$M$ is number of shift registers, which is used in convolutional encoder construction.
$(L + M)$ represents the output of encoder length.

In practice, $L$ is longer than $M$, which means that $M$ length can be neglected in proportion to $L$ length. So the code rate equation can be simplified to Eq. (5.2).

$$R = 1/n \tag{5.2}$$

where $L \geq M$. Eq. (5.2) gives the code rate in case of the input of encoder is one bit as shown in Fig. 5.1, where this figure gives convolutional encoder with code rate $R = 1/2$, $n = 2$.

There is another important term called constraint length $K$ of convolutional code. This term $K$ is used to define the convolutional encoder characteristic. $K$ defined as the number of shifts over a single input bit can influence the encoder output. In case of an encoder with an M-stage shift register, the length of encoder memory equals $M$ bits, the constraint length of this encoder $K = M + 1$. So, the number of shifts which are required for input bit to enter the shift register and finally come out of the encoder is ($K$) the constraint length of the encoder [2].

Figure 5.1 illustrates a convolutional encoder with $n = 2$ and $K = 3$. Hence, the code rate of this encoder is $R = 1/2$. The input of encoder is single bit at a time. This convolutional encoder output represents every input bit by two bits as codeword of the encoder input.

In case of convolutional codes, the input data which is the input of encoder is not divided into blocks as in case of block codes, but the input of convolutional encoder is sequence of data bits. Also, the output of the encoder is not in blocks form as in case of block codes, but it is in series of codeword bits. Convolutional codes can be used for correcting random and burst errors. Convolutional codes are defined by this form,

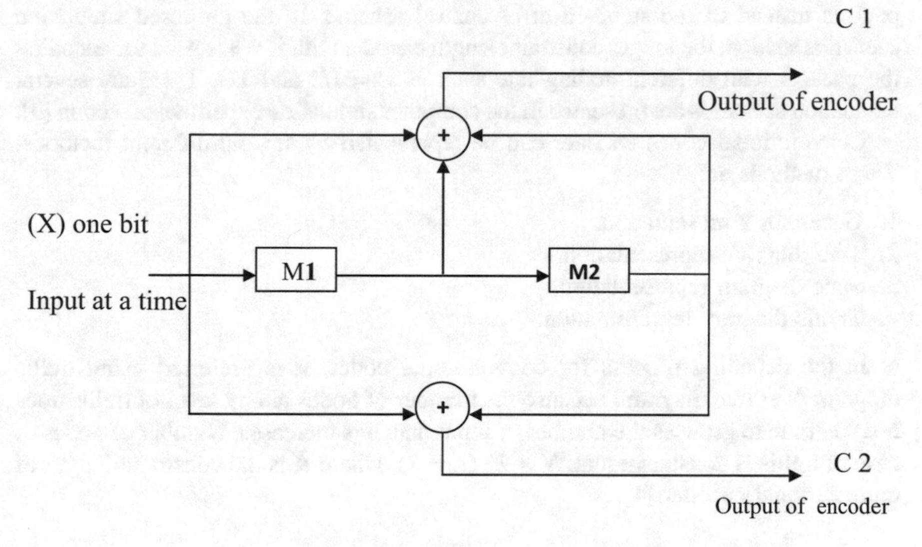


**Fig. 5.1** Example of convolutional encoder, with $k = 1$, $n = 2$, $R = 1/2$, and $K = 3$

{Convolutional code ($k$, $n$, and $K$)}, where $k$ is the number of input data bits, $n$ is the number of output encoded data bits at one time interval, and $K$ is constraint length.

Convolutional codes system contents as shown in Fig. 5.2 firstly consists of the convolutional encoder, its code rate $R = 1/2$, and its constraint length $K = 3$, which are used in the presented computer simulation experiments of WPANs-Bluetooth system. The second block in Fig. 5.2 is the communications channel, the simulation model covers the performance studying of WPANs-Bluetooth system over two communications channels first AWGN channel and Rayleigh-flat fading channel. The last block in this figure is the decoder of convolutional code. There are many decoding methods for convolutional codes. The most common method is Viterbi's likelihood algorithm for digital communication. The maximum likelihood decoding implies selecting a codeword which is closest to the received word. This decoding method is used in the presented simulation model. Viterbi method is very difficult where the decoder is very complex in case of longer constraint length. Constraint length $K = 3$ convolutional encoder is utilized for encoding the payload part of the Bluetooth

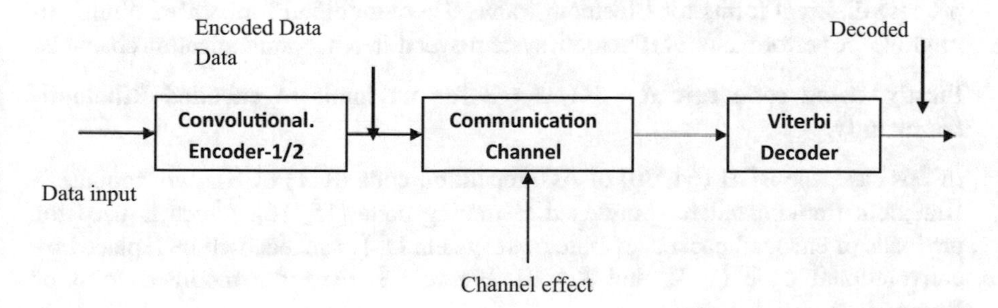


**Fig. 5.2** Convolutional codes system contents

packets instead of the standard error control scheme. In the proposed simulation scenarios section, the longer constraint length encoder with $K = 8$ is used for encoding the packets with different coding rate such as $R$ = 1/2 and 1/3. There are several simulation scenarios are presented in the computer simulation experiments section [3].

Convolutional codes encoder can be represented by several different methods. These methods are

1. Generator representation.
2. Tree diagram representation.
3. State diagram representation.
4. Trellis diagram representation.

In the decoding process for convolutional codes, it is preferred using trellis diagram over tree diagram, because the number of nodes at any level of trellis does not continue to grow as the number of input data bits increases. Number of nodes in case of trellis is constant equal, $N = 2^\wedge (K - 1)$, where $K$ is the constraint length of convolutional encoder [4].

## 5.3 Bluetooth Frame Format with Convolutional Codes

All of the channel codes, which are used in Bluetooth frames, are of block codes type. Performance of Bluetooth system in case of its channel codes is discussed in previous chapters. A new modification, which will happen, is using convolutional code type with Bluetooth frames. This modification improves the performance of Bluetooth system by using convolutional code over AWGN channel and Rayleigh-block fading channel [5].

### *5.3.1 Bluetooth Frame Format Modification ($DM_x$ Packets)*

There are two types of convolutional codes which are used in the modification: convolutional code with rate $R$ = 1/2, $K$ = 3 and $K$ = 8 constraint length and convolutional code with code rate $R$ = 1/3 with $K$ = 3 and $K$ = 8 constraint length. Using new codes will change the length of Bluetooth frame. In the following cases, we will propose different forms for Bluetooth frame. These modified formats are aiming to improve the performance of Bluetooth system over different communication channels.

**Firstly, using code rate $R$ = 1/2, $K$ = 3 for payloads of encoded Bluetooth frame only**.

In this case, the BCH (64, 30) of AC, repetition code (3, 1) of HD will remain as Bluetooth frame standard. Shortened Hamming code (15, 10), which is used for payloads of encoded packets of Bluetooth system $DM_x$ packets, will be replaced by convolutional code (1, 2, and $K$ = 3). Figure 5.3 gives the modified form of Bluetooth frame.

| Access code (AC) | Header (HD) | Payload (PL) |
|---|---|---|
| 64 bits | 54 bits | 0 – 2748 bits |

**Fig. 5.3** Bluetooth frame with convolutional code (1, 2, and $K = 3$) for payload portion

We know the output of convolutional encoder = $n$ ($L$ + $M$), where $M$ is memory of encoder. So, we know maximum length of a payload part = 2 ($DM_x$ + $M$ bits). Maximum length of payload is payload of $DM_5$ packet, which is 2745 bits. In this case, the length of payload will be reduced to keep the length of Bluetooth packet the same standard length. In Bluetooth frames, number of bits per time slot is 366 bits. The maximum length of payload will be 1372 bits.

$$\text{PL length after encoding} = 2(1372 + 2) = 2748 \text{ bits.}$$

**Secondly, using code rate $R = 1/2$, $K = 3$ for three portions of Bluetooth frame**.

In this case, the BCH (64, 30) of AC, repetition code (3, 1) of HD, and code of payloads of encoded packets of Bluetooth system $DM_x$ packets will be replaced by convolutional code (1, 2, and $K = 3$). Figure 5.4 gives the modified form of Bluetooth frame.

We know the output of convolutional encoder = $n$ ($L$ + $M$), where $M$ is memory of encoder. So,

AC length after encoding = 2(30 + 2) = 64.

HD length after encoding = 2(18 + 2) = 40.

PL length will be according to the $DM_x$ type, but the maximum value is 2748. In the following cases, we will neglect M bits number.

**Thirdly, using code rate $R = 1/2$, and $R = 1/3$, $K = 3$ for three portions of Bluetooth frame**.

In this case, the BCH (64, 30) of AC and code of payloads of encoded packets of Bluetooth system $DM_x$ packets will be replaced by convolutional code (1, 2, and $K = 3$). For HD portion, convolutional code (1, 3, and $K = 3$) is used instead of repetition code (3, 1). Figure 5.5 gives the modified form of Bluetooth frame.

| Access code (AC) | Header (HD) | Payload (PL) |
|---|---|---|
| 64 bits | 40 bits | 0 – 2748 bits |

**Fig. 5.4** Bluetooth frame with convolutional code (1, 2, and $K = 3$)

| Access code (**AC**) | Header (**HD**) | Payload (**PL**) |
|---|---|---|
| 60 bits | 54 bits | 0 – 2744 bits |

**Fig. 5.5** Bluetooth frame with convolutional code (1, 2, and $K$ = 3) for AC, PL and (1, 3, and $K$ = 3) for HD

With neglecting the M bits length, the length of each portion is as follows.
AC length after encoding = 60.
HD length after encoding = 54.
PL length will be according to the $DM_x$ type, but its maximum value is 2744.

**Finally, convolutional code (1, 2, and $K$ = 3) for $DM_3$ only**

Using convolutional code (1, 2, and $K$ = 3) for $DM_3$, only the rest $DM_x$ packets, AC, and HD remain the same standard length and error control types. In this case, length of $DM_x$ packet payload before encoding process is 750 bits. After encoding process, its length will be about 1500 bits with neglecting M bits [6].

## 5.3.2 *Bluetooth Frame Format Modification ($DH_x$ Packets)*

**Firstly, using code rate $R$ = 1/2, $K$ = 3 for two portions AC and HD of Bluetooth frame $DH_x$.**

$DH_x$ packets are one of two types of basic Bluetooth packets. These packets are called uncoded packets, where its payloads are sent without FEC. In this modification part using convolutional code (1, 2, and $K$ = 3) instead of BCH code (64, 30) and repetition code (3, 1), these codes are used for AC and HD, respectively. Length of AC and HD after encoding will be 64 bits and 40 bits as shown in second case of $DM_x$ packets.

**Secondly, using code rate $R$ = 1/2, and $R$ = 1/3, $K$ = 3 for two portions of $DH_x$ Bluetooth Frame.**

In this case, we will use two types of convolutional code. First one is convolutional code (1, 2, and $K$ = 3) for AC portion and convolutional code (1, 3, and $K$ = 3) for HD instead of its block codes. Length of AC remains the same as previous modification case, but length of HD is 60 bits without neglecting $M$ bits

## 5.4 General Simulation Assumption

To study the performance of Bluetooth frames over communication different channels and to obtain statistically meaningful results, we should take Bluetooth frames' structure and its specification into our consideration. For these reasons, there are some assumptions in our simulation algorithm as follows [7].

In our simulation, we consider a BPSK modulation instead of GFSK modulation, the latter which is used in modulation of Bluetooth frames.

In our simulation, we use Monte Carlo simulations to evaluate the performance of Bluetooth packets $DM_x$ and $DH_x$. The results of simulation presented consider the case of a BPSK link over channels. In our experiments, we consider a packet of Bluetooth is dropped if there is at least one error after decoding the three portions of Bluetooth frame: the AC, the HD, and the PL, where the last term is payload of encoded packets $DM_x$ [8].

In case of uncoded Bluetooth packets $DH_x$, its packet is discarded if there is single error after decoding the first two portions AC, HD and detecting errors in uncoded PL after receiving.

In Bluetooth system, there are packet types which are sent over multi-time slots and others occupy single time slot that is according to the frequency hopping technique. The presented simulation model covers all the packets of WPAN-Bluetooth system with its different format. Different time slots are occupied by the packet, for example short packet occupies one time slot. WPANs-Blueetooth network can be considered adaptive technology, This WPAN netowrk uses adaptive packet technique principle in the transmission process. Bluetooth frames have different lengths according to its time slots which are occupied by these frames. The channel of simulation is considered time-invariant. That means $\Delta f$ = zero (Doppler spread is equal to zero).

In our simulation, hard decision is assumed at receiver in decoding process for different channel code situations

In our simulation of Bluetooth packets performance with using of convolutional codes as error control codes for payloads (PL) only or for all Bluetooth frame portions, we are using Viterbi's likelihood algorithm in decoding process of encoded data at receiver, where Viterbi algorithm is considered the most common method of convolutional codes decoding [3].

## 5.5 Simulation Results

This section investigates the performance of basic Bluetooth packets ($DM_x$ packets and $DH_x$ packets) over AWGN channels in the case of the proposed coding schemes and compares it to its performance in case of standard coding schemes in Bluetooth systems. Several cases are conducted in our simulation for this purpose [8].

Simulation results of $DM_x$ and $DH_x$ packets in standard form are shown in Figs. 4.4 and 4.5, respectively. These figures are discussed in Chap. 4.

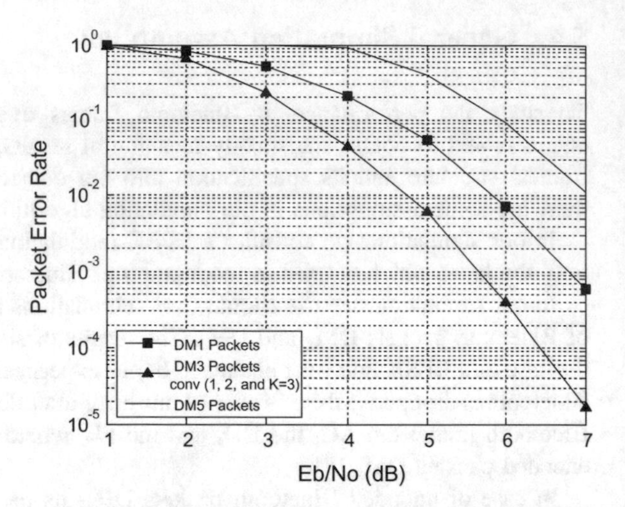


**Fig. 5.6** Simulated PER of $DM_x$ packets over AWGN channel by using convolutional code (1, 2, and $K = 3$) for PL of $DM_3$ only, the rest as standard case

Figure 5.6 shows simulation result of $DM_x$ packets performance over AWGN channel in case of using convolutional code ($k = 1$, $n = 2$, and $K = 3$) for $DM_3$ payloads only. This figure reveals the superiority of convolutional code for the AWGN channel. Performance of $DM_3$ packets in case of convolutional code is better than $DM_1$ in standard case [shortened Hamming code (15, 10)]. As shown in this figure by using convolutional. At PER $10^{-3}$ the values of SNR are 5.6, 6.8, and above 7 dB for $DM_3$ with using convolutional code (1, 2, and $K = 3$), $DM_1$, and $DM_5$ in standard case, respectively.

Simulation experiment results of $DM_x$ packets performance evaluation over Rayleigh-flat fading channel are given in Fig. 5.7. In the presented experiment, the packets are encoded using standard error control scheme except DM3 packets are encoded by the convolutional code (k=1, n=2, K=3). This experiment results reveal that the proposed channel coding scheme improves the performance of the encoded $DM_3$ packets. In this simulation case, the proposed encoded $DM_3$ performs better than the standard short encoded $DM_1$ by about 1.5 dB.

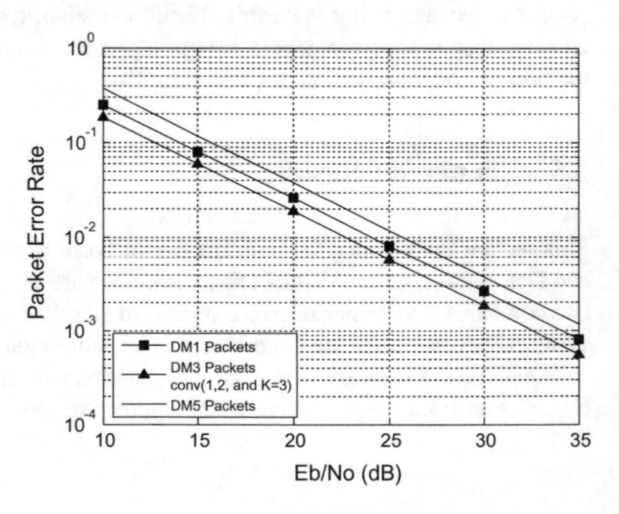


**Fig. 5.7** Simulated PER of $DM_x$ packets on Rayleigh-flat fading channel by using convolutional code (1, 2, and $K = 3$) for PL of $DM_3$ only, the rest as standard case

In standard case of $DM_x$ Bluetooth Packets, performance of $DM_3$ and $DM_5$ is closed. With new proposed form of $DM_3$, it performs better than $DM_5$ by 3.5 dB.

### 5.5.1 AWGN Channel

Figure 5.8 shows simulation result of $DM_x$ packets performance over AWGN channel in case of using convolutional code ($k = 1$, $n = 2$, and $K = 3$) for $DM_x$ payloads only. This figure reveals the superiority of convolutional code for the AWGN channel. Performance of $DM_x$ packets in case of convolutional code is better than shortened Hamming code (15, 10).

The performance gain of $DM_x$ packet is 2.5 dB over $DM_x$ packets [shortened Hamming code (15, 10)]. As shown in Fig. 5.8, at PER $10^{-3}$, the values of SNR are 4.8, 5.8, and 6.1 dB for $DM_1$, $DM_3$, and $DM_5$ with using convolutional code (1, 2, and $K = 3$), respectively.

The following simulation case shows the effect of changing error control codes of AC and HD portions on a performance of $DM_x$ packets over AWGN channel. Simulation results are shown in Fig. 5.9. These results show the performance of the WPAN-Bluetooth network encoded packets with utilizing the convolutional code (1, 2, $K = 3$) for encoding the AC, HD and payload sections instead of the standard error control schemes. As shown in this figure, at PER $10^{-3}$, the values of SNR are 4.4, 5.6, and 5.7 dB for $DM_1$, $DM_3$, and $DM_5$ with using convolutional code (1, 2, and $K = 3$), respectively.

The results which are shown in Figs. 5.8 and 5.9 are very close, which means that the using of convolutional code in AC, HD is ineffective on AWGN, where the performance gain is less than 0.5 dB.

The following simulation result shows effect of constraint length ($K$) of convolutional encoder on $DM_x$ packets performance over AWGN channel. Figure 5.10

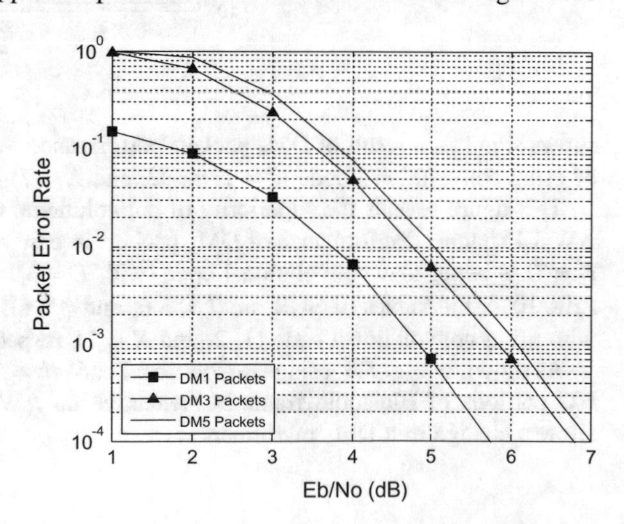


**Fig. 5.8** Simulated PER of $DM_x$ packets over AWGN channel by using convolutional code (1, 2, and $K = 3$) for PL only

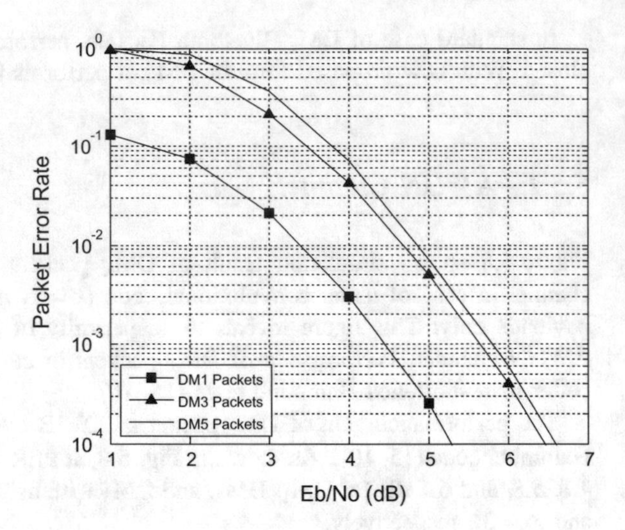


**Fig. 5.9** Simulated PER of $DM_x$ packets over AWGN channel by using convolutional code (1, 2, and $K$ = 3) for AC, HD, and PL

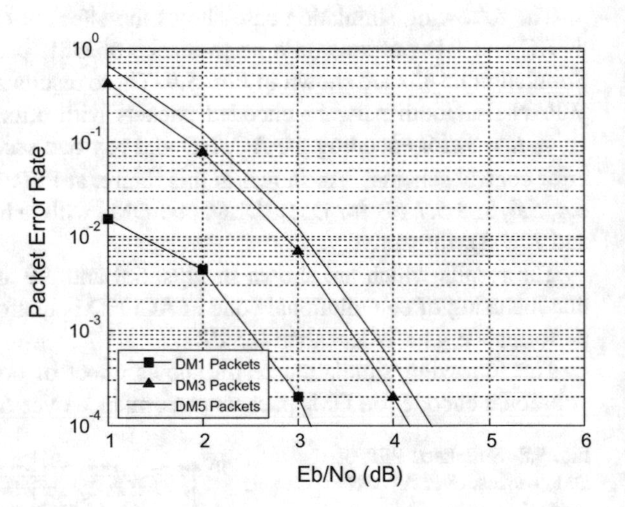


**Fig. 5.10** Simulated PER of $DM_x$ packets over AWGN channel by using convolutional code (1, 2, and $K$ = 7) for PL only

shows simulation result of $DM_x$ packets performance over AWGN channel in case of using convolutional code ($k = 1$, $n = 2$, and $K = 7$) for $DM_x$ payloads only.

This figure reveals the superiority of convolutional code with longer $K$ for the AWGN channel. Performance of $DM_x$ packets in case of convolutional code (with $K = 7$) is better than convolutional code (with $K = 3$). As shown in this figure, at PER $10^{-3}$, the values of SNR are 2.5, 3.6, and 3.8 dB for $DM_1$, $DM_3$, and $DM_5$ with using convolutional code (1, 2, and $K = 7$), respectively [9].

As shown in Fig. 5.9, using convolutional code ($k = 1$, $n = 2$, and $K = 3$) in AC, HD portions of Bluetooth frame is ineffective on AWGN channel and does not allow a change in a $DM_x$ performance.

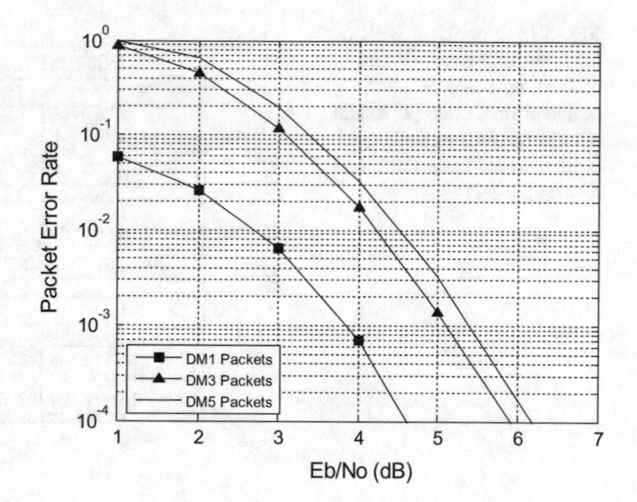


**Fig. 5.11** Simulated PER of DM packets over AWGN channel by using convolutional code (1, 2, and $K = 3$) for AC, payload, and convolutional code (1, 3, and $K = 3$) for HD part

Figure 5.11 gives simulation result of $DM_x$ performance over AWGN channel with using convolutional code ($k = 1$, $n = 2$, and $K = 3$) for $DM_x$ payloads and AC portions only; HD portion is encoded by convolutional code ($k = 1$, $n = 3$, and $K = 3$).

In this figure at PER $10^{-3}$, the values of SNR are 3.8, 5.1, and 5.4 dB for $DM_1$, $DM_3$, and $DM_5$, respectively. Figures 5.8, 5.9, and 5.11 reveal that changing FEC of AC and HD is ineffective and does not give observable improving in $DM_x$ performance.

The Bit Error Rate (BER) metric of $DM_x$ packets over AWGN channel with using convolutional code ($k = 1$, $n = 2$, and $K = 3$) for $DM_x$ payloads and AC portions encoding. Also, the HD portion is encoded by convolutional code ($k = 1$, $n = 3$, and $K = 3$) is shown in Fig. 5.12.

**DHx packets over AWGN channel with using convolutional code in AC and HD portions.**

In the following simulation cases, we will study the effect of using convolutional code on the $DH_x$ packets performance over AWGN channel. There are two cases: first case by using convolutional code ($k = 1$, $n = 2$, and $K = 3$) for AC and HD and second case using previous code for AC only and using convolutional code $R = 1/3$ with $K = 3$ for HD portion plus uncoded payloads [1].

Figure 5.13 shows the simulation result of $DH_x$ packets over AWGN with using convolutional code ($k = 1$, $n = 2$, and $K = 3$) for AC and HD portions with uncoded payloads. As shown in this figure, using this code gives little improvement in performance of $DH_x$ packets comparing to $DH_x$ packets performance in standard case. It appears that using this code in AC and HD is not efficient.

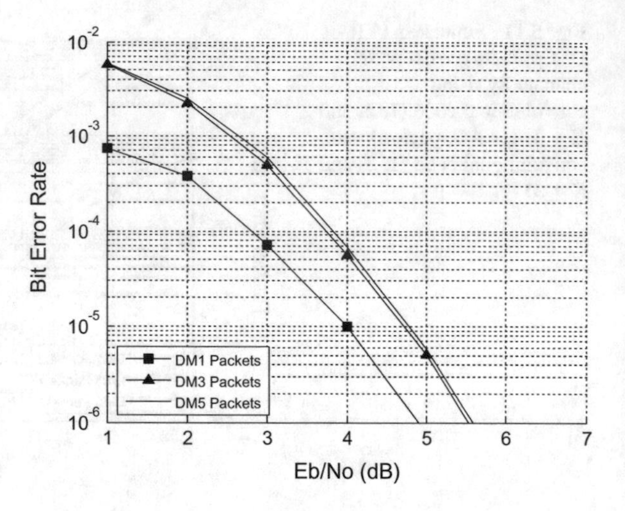


**Fig. 5.12** Simulated BER of $DM_x$ packets over AWGN channel by using convolutional code (1, 2, and $K = 3$) for AC, payload, and convolutional code (1, 3, and $K = 3$) for HD

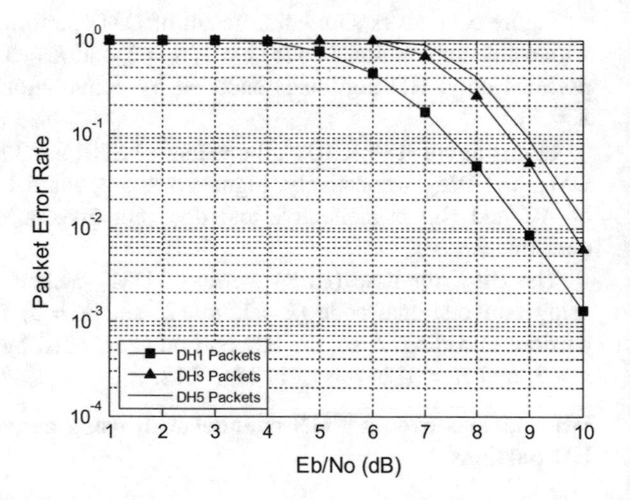


**Fig. 5.13** Simulated PER of $DH_x$ packets in case of AWGN channel by using convolutional code (1, 2, and $K = 3$) for AC and HD portions, uncoded payload

Second computer simulation experiment of the uncoded $DH_x$ packets using the convolutional code for encoding the AC and HD portions only utilizing different code rate, in this experiment, the code rate is 1/3. This experiment is carried out to evaluate the performance of the WPAN-Bluetooth packets with different error control schemes for packet header encoding. Simulation result of this case is shown in Fig. 5.14. As shown in this figure, the $DH_x$ performance is not improved more than 0.2 dB.

Simulation result of BER of $DH_x$ packets is shown in Fig. 5.15 over AWGN channel with using the same code type as previous figure. This figure reveals that $DH_x$ performance over AWGN channel is nearly as standard case.

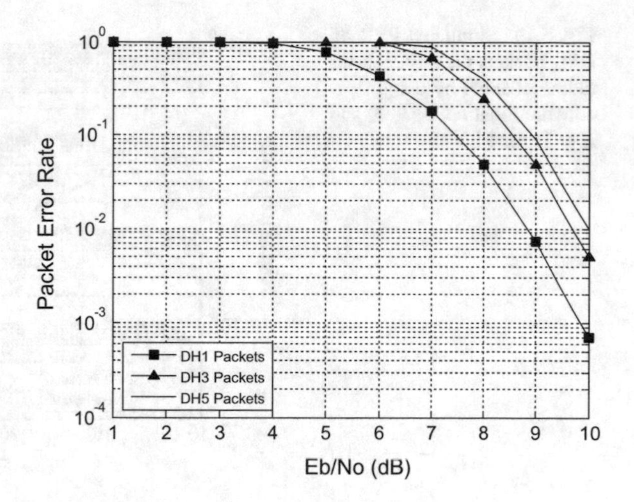


**Fig. 5.14** Simulated PER of $DH_x$ packets in case of AWGN channel by using convolutional code (1, 2, and $K = 3$) for AC and convolutional code (1, 3, and $K = 3$) for HD portion

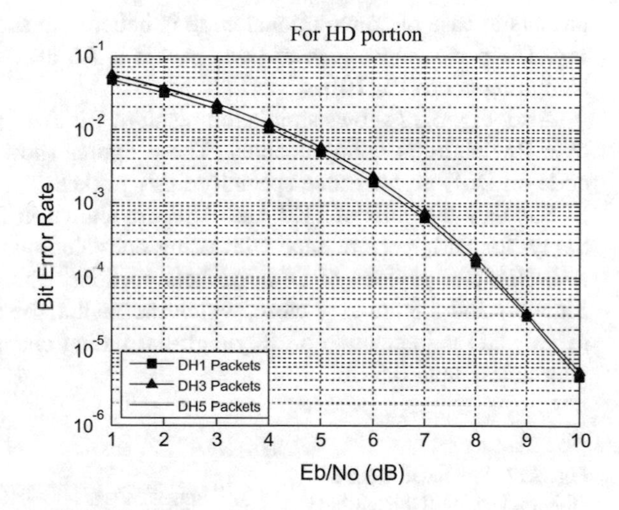


**Fig. 5.15** Simulated BER of $DH_x$ packets in case of AWGN channel by using convolutional code (1, 2, and $K = 3$) for AC, payload, and convolutional code (1, 3, and $K = 3$) for HD

## 5.5.2 *Rayleigh-Flat Fading Channel*

This section studies the performance of basic Bluetooth packets ($DM_x$ packets and $DH_x$ packets) over Rayleigh-flat fading channel in the case of the proposed coding schemes and compares it to its performance in case of standard coding schemes in Bluetooth systems. Several cases are conducted in our simulation for this purpose [10].

Figure 5.16 shows simulation result of Packet Error Rate (PER) of $DM_x$ packets performance over Rayleigh-flat fading channel in case of using convolutional code ($k = 1$, $n = 2$, and $K = 3$) for $DM_x$ payloads only. This figure reveals the superiority of convolutional code for the Rayleigh-flat fading channel. Performance of $DM_x$

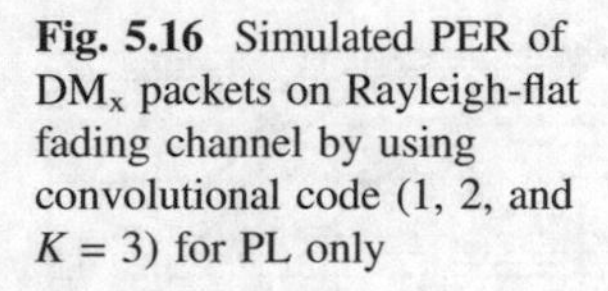

**Fig. 5.16** Simulated PER of $DM_x$ packets on Rayleigh-flat fading channel by using convolutional code (1, 2, and $K$ = 3) for PL only

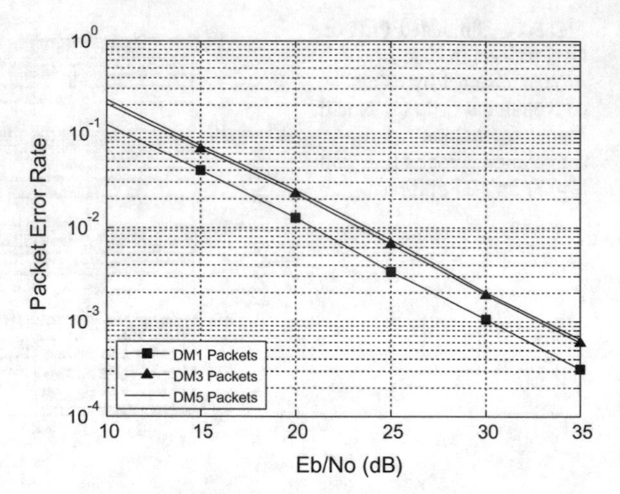


packets in case of convolutional code is better than shortened Hamming code (15, 10). The performance gain of $DM_x$ packet is 2.5 dB over $DM_x$ packets [shortened Hamming code (15, 10)].

Also, Fig. 5.17 shows simulation result of Bit Error Rate (BER) of $DM_x$ packets over Rayleigh-flat fading channel. These figures show the effect of convolutional code on $DM_x$ performance specially $DM_1$ packet.

Simulation results of PER and BER are shown in Figs. 5.18 and 5.19, respectively, for $DM_x$ performance with using convolutional code (1, 2, and $K$ = 3) for AC, HD, and payload of $DM_x$ packets portions. The results which are shown in Figs. 5.7 and 5.8 are very close, which means that the using of convolutional code in AC, HD is ineffective on Rayleigh-flat fading channel, where the performance gain is less than 0.5 dB.

**Fig. 5.17** Simulated BER of $DM_x$ packets on Rayleigh-flat fading channel by using convolutional code (1, 2, and $K$ = 3) for PL

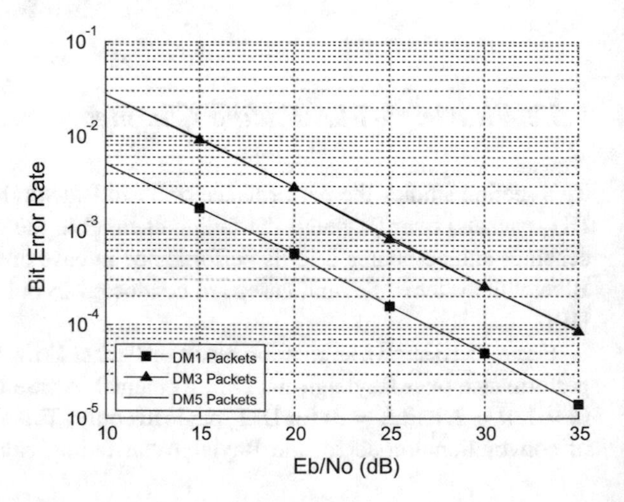

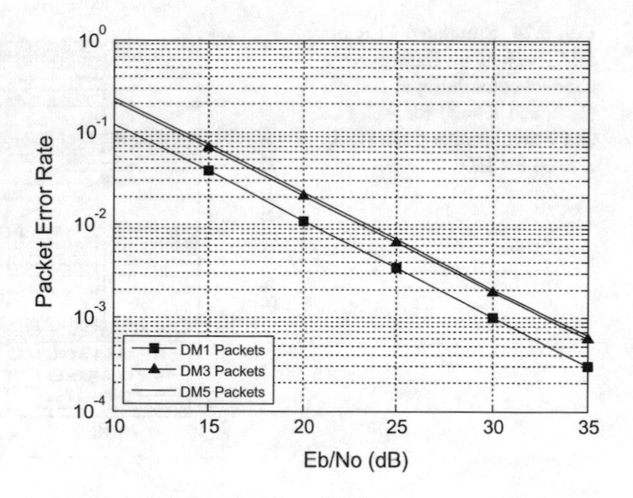


**Fig. 5.18** Simulated PER of $DM_x$ packets on Rayleigh-flat fading channel by using convolutional code (1, 2, and $K = 3$) for AC, HD, and PL

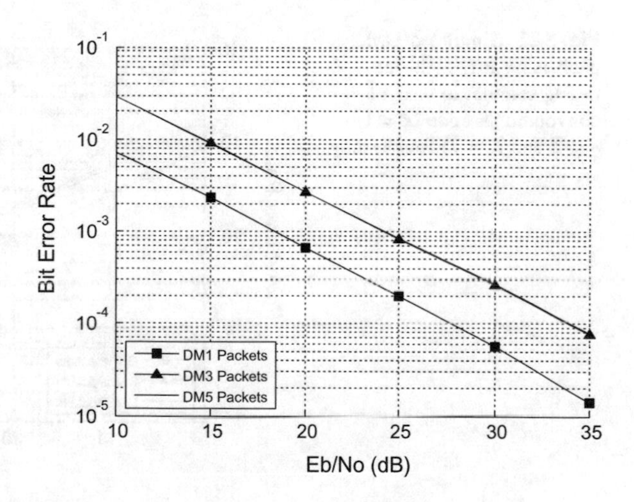


**Fig. 5.19** Simulated BER of $DM_x$ packets on Rayleigh-flat fading channel by using convolutional code (1, 2, and $K = 3$) for AC, HD, and PL

Figures 5.16, 5.17, 5.18, and 5.19 reveal that the using of convolutional code in AC, HD is ineffective on Rayleigh-flat fading channel, where the performance gain is less than 0.5 dB. Using convolutional code ($k = 1$, $n = 2$, and $K = 3$) in HD portion will decrease the length of Bluetooth frame as described in previous section.

Figure 5.20 gives the simulation result of PER for $DM_x$ packets over the same channel and the same coding schemes but using convolutional code ($k = 1$, $n = 3$, and $K = 3$) in HD portion encoding. This figure reveals that using this code with HD is ineffective and does not allow a change in a $DM_x$ performance.

Fig. 5.21 shows the simulation result of PER of $DM_x$ packets over Rayleigh-flat fading channel with using convolutional code ($k = 1$, $n = 2$, and $K = 7$).

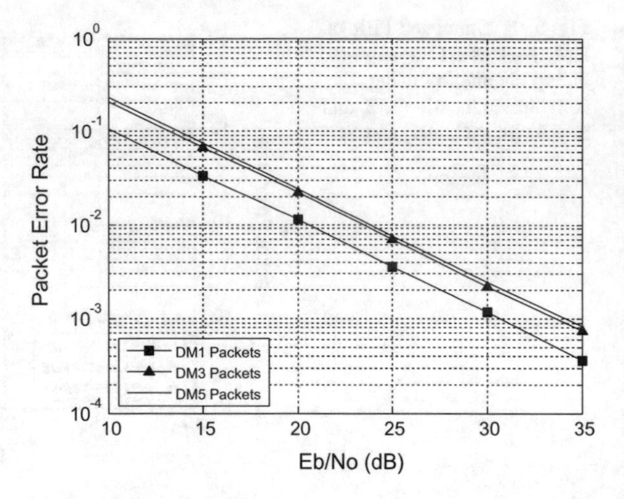


**Fig. 5.20** Simulated PER of $DM_x$ packets on fading channel convolutional code (1, 2, and $K = 3$) for AC, PL, and convolutional code (1, 3, $K = 3$) for HD

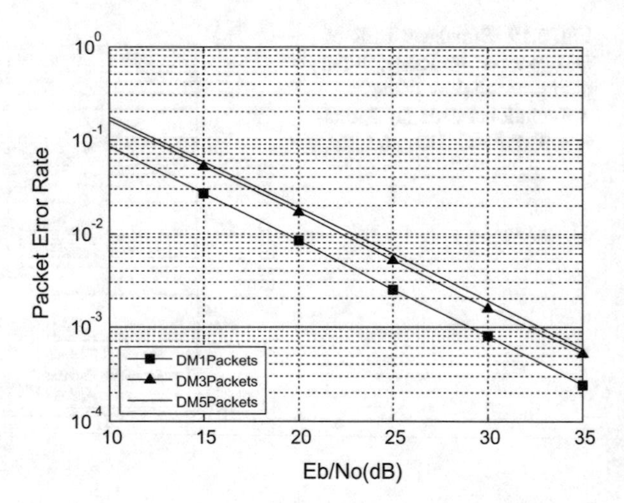


**Fig. 5.21** Simulated $DM_x$ packets on Rayleigh-flat fading channel in case of convolutional code ($k = 1$, $n = 2$ and $K = 7$) for PL only

This simulation case studies the effect of longer constraint length of convolutional encoder ($K$) on the $DM_x$ packets performance. As shown this figure, $DM_x$ packets perform better with longer $K$.

From Figs. 5.18, 5.20, and 5.21, we can conclude that using convolutional code with longer $K$ of its encoder (for PL only) is better than using this code for each one portions (AC, HD, and PL) of Bluetooth frame, where the first gives a good change in the $DM_x$ performance especially for shorter packets $DM_1$ compared with $DM_x$ and $DH_x$ packets performance in standard case.

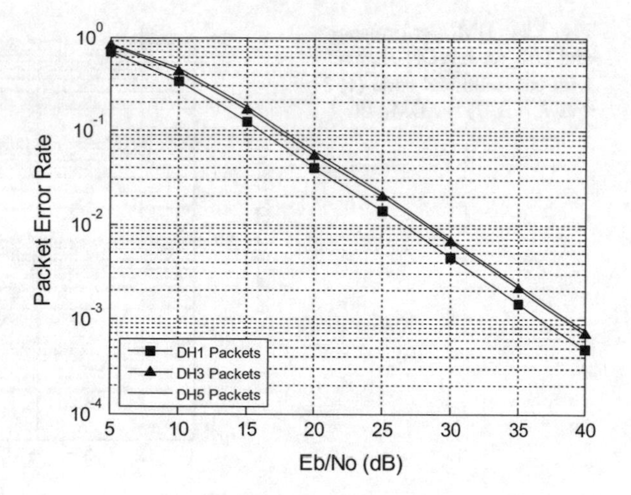


**Fig. 5.22** Simulated PER of $DH_x$ packets on Rayleigh-flat fading channel by using convolutional code (1, 2, and $K$ = 3) for AC and HD

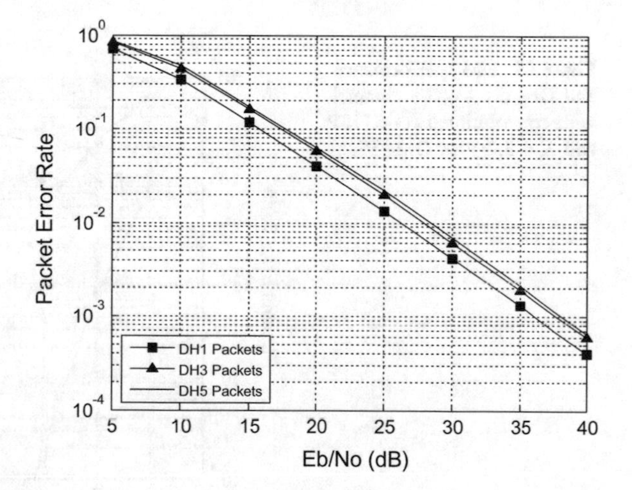


**Fig. 5.23** Simulated PER of $DH_x$ packets on Rayleigh-flat fading channel by using convolutional code (1, 2, and $K$ = 3) for AC and convolutional code (1, 3, and $K$ = 3) for HD

**$DH_x$ packets over Rayleigh-flat fading channel with using convolutional code in AC and HD portions.**

Figure 5.22 shows the simulation result of $DH_x$ packets over Rayleigh-flat fading channel with using convolutional code ($k$ = 1, $n$ = 2, and $K$ = 3) for AC and HD portions with uncoded payloads. As shown in this figure, using this code gives little improvement in performance of $DH_x$ packets comparing to $DH_x$ packets performance in standard case.

Another case of $DH_x$ packets simulation is presented in this experiment using the convolutional code in AC portion only and using code $R$ = 1/3 with the constraint length $K$ = 3 for encoding the HD portion. Simulation result of this case is shown in Fig. 5.23. As shown in this figure, the $DH_x$ performance is not improved more than 0.5 dB.

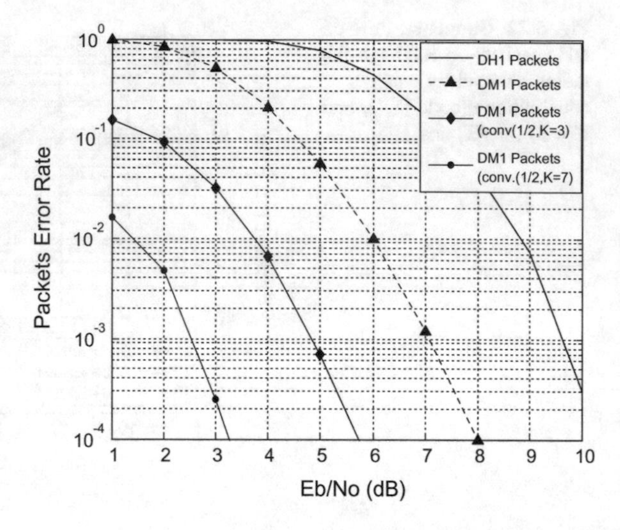


**Fig. 5.24** $DM_1$ performance and $DH_1$ on AWGN channel with convolutional code (1, 2 and $K$ = 3, 7) for $DM_1$ PL

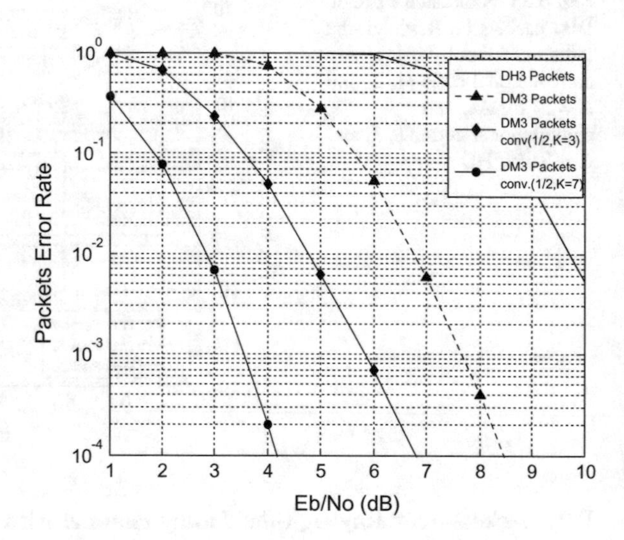


**Fig. 5.25** $DM_3$ performance and $DH_3$ on AWGN channel with convolutional code (1, 2 and $K$ = 3, 7) for $DM_3$ PL

## 5.6 Performance Comparison

This section presents group of figures. These figures as will be discussed show the performance of ACL Bluetooth packets with convolutional codes. There are two groups of figures. First group compromised from Figs. 5.24, 5.25, and 5.26 shows the performance of ACL Bluetooth packets over AWGN channel. Second group contains Figs. 5.27, 5.28, and 5.29 which show the packets performance over Rayleigh-flat fading channel.

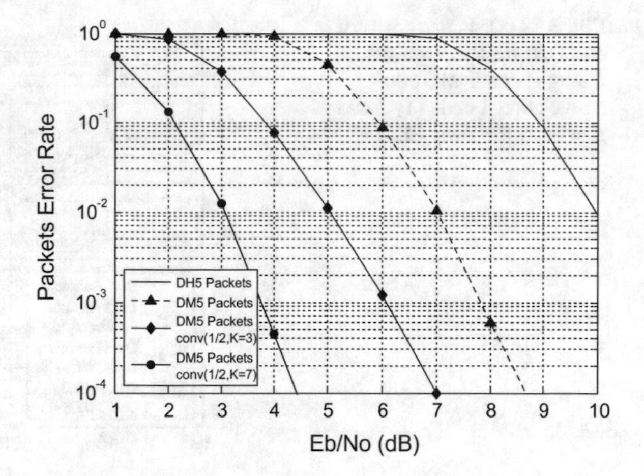


**Fig. 5.26** $DM_5$ performance and $DH_5$ on AWGN channel with convolutional code (1, 2 and $K = 3$, 7) for $DM_5$ PL

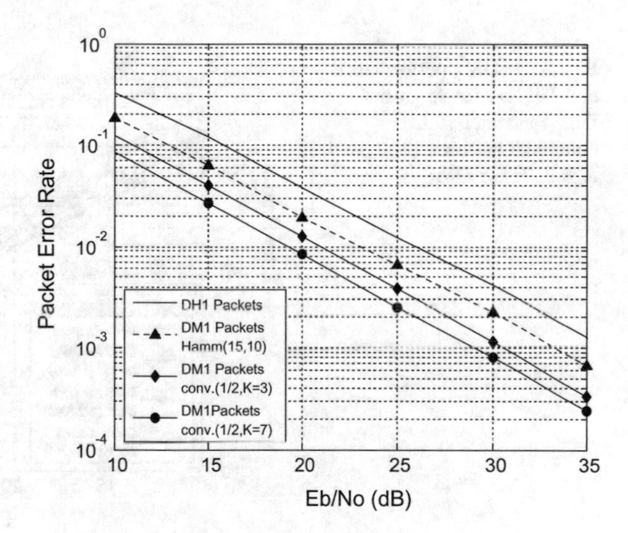


**Fig. 5.27** $DM_1$ performance and $DH_1$ on Rayleigh-flat fading channel with convolutional code (1, 2 and $K = 3$, 7) for $DM_1$ PL

## AWGN Channel Group

Figure 5.24 compares the performance of uncoded $DH_1$ packets with the performance of encoded $DM_1$ packets (both standard case and convolutional code with different constraint length $K$ values) that is over AWGN channel. As shown in this figure by using convolutional code, $DM_1$ performs better than $DH_1$ and $DM_1$ with shortened Hamming code. Convolutional code with $K = 7$ gives performance better than $K = 3$. At PER $10^{-3}$, the values of SNR are 2.5, 4.8, 7, and 9.7 dB for convolutional code (1, 2, and $K = 7$), convolutional code (1, 2, and $K = 3$), shortened Hamming code (15, 10) of $DM_1$ packets, and uncoded $DH_1$ packets, respectively, as shown in Table 5.1.

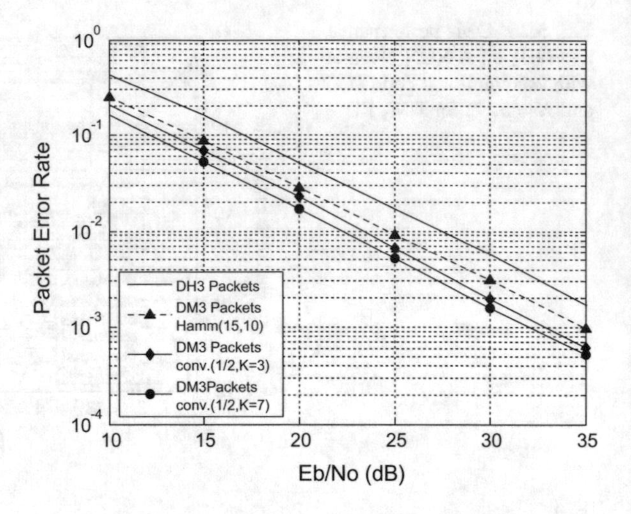


**Fig. 5.28** $DM_3$ performance and $DH_3$ on Rayleigh-flat fading channel with convolutional code (1, 2 and $K$ = 3, 7) for $DM_3$ PL

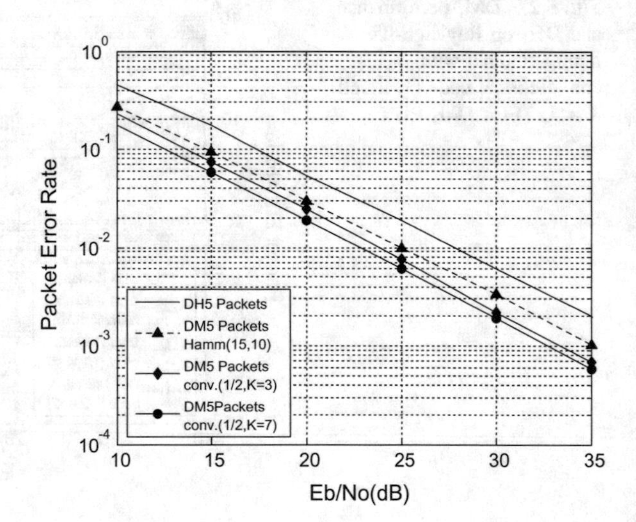


**Fig. 5.29** $DM_5$ performance and $DH_5$ on Rayleigh-flat fading channel with convolutional code (1, 2, and $K$ = 3, 7) for $DM_5$

**Table 5.1** $DM_1$ and $DH_1$ performance comparison over AWGN channel with standard and proposed case

| PER | SNR (dB) of $DM_1$ and $DH_1$ of Bluetooth ACL packets | | | |
|---|---|---|---|---|
| | $DH_1$ standard | $DM_1$ standard | $DM_1$ (Conv. $R = 1/2$) $K = 3$ | $DM_1$ (Conv. $R = 1/2$) $K = 7$ |
| $10^{-3}$ | 9.66 | 7.1 | 4.8 | 2.5 |
| $10^{-4}$ | >10 | 8 | 5.6 | 3.3 |

**Table 5.2** $DM_3$ and $DH_3$ performance comparison over AWGN channel with standard and proposed case

| PER | SNR (dB) of $DM_3$ and $DH_3$ of Bluetooth ACL packets | | | |
|---|---|---|---|---|
| | $DH_3$ standard | $DM_3$ standard | $DM_3$ (Conv. $R = 1/2$) $K = 3$ | $DM_3$ (Conv. $R = 1/2$) $K = 7$ |
| $10^{-3}$ | >10 | 7.66 | 5.77 | 3.44 |
| $10^{-4}$ | >10 | 8.4 | 6.8 | 4.1 |

**Table 5.3** $DM_5$ and $DH_5$ performance comparison over AWGN channel with standard and proposed case

| PER | SNR (dB) of $DM_5$ and $DH_5$ of Bluetooth ACL packets | | | |
|---|---|---|---|---|
| | $DH_5$ standard | $DM_5$ standard | $DM_5$ (Conv. $R = 1/2$) $K = 3$ | $DM_5$ (Conv. $R = 1/2$) $K = 7$ |
| $10^{-3}$ | >10 | 7.85 | 6.1 | 3.8 |
| $10^{-4}$ | >10 | 8.6 | 7 | 4.35 |

Figure 5.25 gives comparison between the performance of $DM_3$ packets (both standard case and convolutional code with different constraint length $K$ values) and $DH_3$ packets that are over AWGN channel. $DM_3$ packets perform better than $DH_3$. The best performance of $DM_3$ is given by using convolutional code with $K = 7$. At PER $10^{-3}$, the values of SNR are 3.5, 5.8, 7.6, and over 10 dB for convolutional code (1, 2, and $K = 7$), convolutional code (1, 2, and $K = 3$), shortened Hamming code (15, 10) of $DM_3$ packets, and uncoded $DH_3$ packets, respectively, as shown in Table 5.2.

As shown in Fig. 5.25, performance of $DM_1$ and $DH_1$ is better than $DM_3$ and $DH_3$, respectively.

In the same manner in case of $DM_5$ (the same coding schemes) and $DH_5$, the result of performance comparison is shown in Fig. 5.26. These figures reveal that the performance of $DM_5$ and $DM_3$ are very close.

At PER $10^{-3}$, the values of SNR are 3.7, 6.1, 7.8, and over 10 dB for convolutional code (1, 2, and $K = 7$), convolutional code (1, 2, and $K = 3$), shortened Hamming code (15, 10) of $DM_5$ packets, and uncoded $DH_5$ packets, respectively, as shown in Table 5.3.

**Table 5.4** $DM_1$ and $DH_1$ performance comparison over Rayleigh-flat fading channel with standard and proposed case

| PER | SNR (dB) of $DM_1$ and $DH_1$ of Bluetooth ACL packets | | | |
|---|---|---|---|---|
| | $DH_1$ standard | $DM_1$ standard | $DM_1$ (Conv. $R = 1/2$) $K = 3$ | $DM_1$ (Conv. $R = 1/2$) $K = 7$ |
| $10^{-2}$ | 26.17 | 23.23 | 20.88 | 19.11 |
| $10^{-3}$ | >35 | 33.3 | 30.5 | 28.5 |

These previous results reveal that the convolutional code gives a good difference in performance for $DM_x$ packets on AWGN channel as compared to $DH_x$ packets performance over the same channel, as shown in previous Figs. 5.24, 5.25, and 5.26.

**Rayleigh-Flat Fading Channel Group**

Figure 5.27 compares the performance of uncoded $DH_1$ packets with the performance of encoded $DM_1$ packets (both standard case and convolutional code with different constraint length $K$ values) that is over Rayleigh-flat fading channel. As shown in this figure by using convolutional code, $DM_1$ performs better than $DH_1$ and $DM_1$ with shortened Hamming code. Convolutional code with $K = 7$ gives performance better than $K = 3$. At PER $10^{-3}$, the values of SNR are 28.5, 30.5, 33.3, and over 35 dB for convolutional code (1, 2, and $K = 7$), convolutional code (1, 2, and $K = 3$), shortened Hamming code (15, 10) of $DM_1$ packets, and uncoded $DH_1$ packets, respectively, as shown in Table 5.4.

Figure 5.28 gives comparison between the performance of $DM_3$ packets (both standard case and convolutional code with different constraint length $K$ values) and $DH_3$ packets that are over Rayleigh-flat fading channel. $DM_3$ packets perform better than $DH_3$. The best performance of $DM_3$ is given by using convolutional code with $K = 7$. At PER $10^{-3}$, the values of SNR are 31.87, 32.8, 34.8, and over 35 dB for convolutional code (1, 2, and $K = 7$), convolutional code (1, 2, and $K = 3$), shortened Hamming code (15, 10) of $DM_3$ packets, and uncoded $DH_3$ packets, respectively. Performance of $DM_1$ and $DH_1$ is better than $DM_3$ and $DH_3$, respectively, as shown in Table 5.5.

**Table 5.5** $DM_3$ and $DH_3$ performance comparison over Rayleigh-flat fading channel with standard and proposed case

| PER | SNR (dB) of $DM_3$ and $DH_3$ of Bluetooth ACL packets | | | |
|---|---|---|---|---|
| | $DH_3$ standard | $DM_3$ standard | $DM_3$ (Conv. $R = 1/2$) $K = 3$ | $DM_3$ (Conv. $R = 1/2$) $K = 7$ |
| $10^{-2}$ | 27.64 | 24.8 | 23.24 | 22.35 |
| $10^{-3}$ | >35 | 34.8 | 32.8 | 31.87 |

**Table 5.6** $DM_5$ and $DH_5$ performance comparison over Rayleigh-flat fading channel with standard and proposed case

| PER | SNR (dB) of $DM_5$ and $DH_5$ of Bluetooth ACL packets | | | |
|---|---|---|---|---|
| | $DH_5$ standard | $DM_5$ standard | $DM_5$ (Conv. $R = 1/2$) $K = 3$ | $DM_5$ (Conv. $R = 1/2$) $K = 7$ |
| $10^{-2}$ | 27.6 | 25 | 23.8 | 22.9 |
| $10^{-3}$ | >35 | 35 | 33 | 32.5 |

In the same manner in case of $DM_5$ (the same coding schemes) and $DH_5$, the result of performance comparison is shown in Fig. 5.29. These figures reveal that the performance of $DM_5$ and $DM_3$ are very close. As shown in these figures, using the convolutional code in Bluetooth packets has a good effect on shorter Bluetooth packets ($DM_1$) performance. This effect is little on longer packets $DM_3$. In case of longest Bluetooth packets $DM_5$, convolutional code is ineffective on Rayleigh fading channel.

As shown in Fig. 5.29, at PER $10^{-3}$, the values of SNR are 32.5, 33, 35, and over 35 dB for convolutional code (1, 2, and $K = 7$), convolutional code (1, 2, and $K = 3$), shortened Hamming code (15, 10) of $DM_5$ packets, and uncoded $DH_5$ packets, respectively, as shown in Table 5.6.

These previous results reveal that the convolutional code gives a good difference in performance for $DM_x$ packets on Rayleigh-flat fading channel compared to $DH_x$ packets performance over the same channel, as shown in previous Figs. 5.27, 5.28, and 5.29. Convolutional code has good effect on performance of shorter Bluetooth packets ($DM_1$), but this effect is a little on longer Bluetooth packets $DM_3$ and $DM_5$. Effects on convolutional code on $DM_5$ packets performance is less than its effect on $DM_1$ on Rayleigh-flat fading channel [10].

**The MATLAB codes of the convolutional codes simulations:**
**The main code of the encoded packets** [11, 12]:

```
%########################################################

clc
close all
clear all

 %_-_-_______________________BLUEOOTH SIMULATION_______________-_-_-_-%

 %THE LENGTH OF BT_PACKET  BEFORE ENCODING PROCESS
 % LENGTH=4+(24+6)+4+18+PAYLOAD
                                  %MEDIUM RATE(K)
 %SO THE LENGTH OF  ====> DM1=56+PAY(160) =216
===========>FEC(15,10)HAM.
 %SO THE LENGTH OF  ====> DM3=56+PAY(1000)=1056
===========>FEC(15,10)HAM.
 %SO THE LENGTH OF  ====> DM5=56+PAY(1830)=1886
===========>FEC(15,10)HAM.
                        %____________HIGH RATE(K)_______________%

 %SO THE LENGTH OF  ====> DH1=56+PAY(240) =296  ===========> NO FEC
 %SO THE LENGTH OF  ====> DH3=56+PAY(1500)=1556 ===========> NO FEC
 %SO THE LENGTH OF  ====> DH5=56+PAY(2745)=2801 ===========> NO FEC
                        %____________AFTER ENCODING__________%
 % THE LENGTH OF BT_PACKET AFTER ENCODING PROCESS
 %LENGTH= 4+64(SYNC WORD)+4+54(HEADER)+PAYLOAD
                                    % MEDIUM RATE(N)
 %SO THE LENGTH OF  ====> DM1=126+PAY(240) =366
<===========FEC(15,10)HAM.
 %SO THE LENGTH OF  ====> DM3=126+PAY(1500)=1626
<===========FEC(15,10)HAM.
 %SO THE LENGTH OF  ====> DM5=126+PAY(2745)=2871
<===========FEC(15,10)HAM.
                          %______________HIGH RATE(N)__________________%

 %SO THE LENGTH OF  ====> DH1=126+PAY(240) =366  ===========> NO FEC
 %SO THE LENGTH OF  ====> DH3=126+PAY(1500)=1626 ===========> NO FEC
 %SO THE LENGTH OF  ====> DH5=126+PAY(2745)=2871 ===========> NO FEC
 % THE RATE WILL BE AS SHOWN

                        % DM1  -------> R1=K1/N1=216/366.
                        % DM3  -------> R3=K3/N3=1056/1626.
                        % DM5  -------> R5=K5/N5=1886/2871.
                        % DH1---------> RH1=K'1/N'1=296/366.
                        % DH3  -------> RH3=K'3/N'3=1556/1626.
                        % DH5  -------> RH5=K'5/N'5=2801/2871.

    %HERE WE KNOW RATE OF EACH BLUETOOTH PACKET TRYING TO CHANGE THE
    %FEC AND COMPARE 1-FIRST USE BLOCK CODES
    %                2-SECOND USE CONV.ECODER WITH PUNCHERING PROCESS
    %                      TO KEEP THE RATE AS MUCH AS WE CAN
    %################### INPUTS DEFINITIONS########################
    %###################                  ########################
    %###################  ACCESS CODE CONV ########################
     STATE_N_AC  =64;%input('STATE NO      STATE_N_DM1  =: ');
```

```
 HALF_ST_AC  =32;%input('1/2STATE NO          HALF_ST_DM1  =: ');
 IN_AC       =1;%input('encoder input         IN_AC   =: ');
 N_out_AC    =2;%input('encoder output        N_ont_AC =: ');
 K_AC        =7;%input('the constraint len.      K_AC =: ');
 L_AC        =30;%input('data length              L_AC =: ');
 M_AC        =6;%input('TAIL length              L_DM1 =: ');
%################################################################
%####################   HEADER PART CONV    ######################
 STATE_N_HD   =64;%input('STATE NO        STATE_N_DM1   =: ');
 HALF_ST_HD   =32;%input('1/2STATE NO        HALF_ST_DM1   =: ');
 IN_HD        =1;%input('encoder input      IN_HD   =: ');
 N_out_HD     =2;%input('encoder output      N_out_HD =: ');
 K_HD         =7;%input('the constraint len.      K_HD =: ');
 L_HD         =18;%input('data length               L_HD =: ');
 M_HD         =6;%input('TAIL length              L_DM1 =: ');
%#################################################################
%####################     DM1  CONV        #######################
STATE_N_DM1 =64;%input('STATE NO        STATE_N_DM1   =: ');
HALF_ST_DM1 =32;%input('1/2STATE NO        HALF_ST_DM1   =: ');
IN_DM1      =1;%input('encoder input      IN_DM1   =: ');
N_out_DM1   =2;%input('encoder output      N_out_DM1 =: ');
K_DM1       =7;%input('the constraint len.      K_DM1 =: ');
L_DM1       =120;%input('data length              L_DM1 =: ');
M_DM1       =6;%input('TAIL length              L_DM1 =: ');
%#################################################################
%####################     DM3  CONV        #######################
STATE_N_DM3 =64;%input('STATE NO        STATE_N_DM3   =: ');
HALF_ST_DM3 =32;%input('1/2STATE NO        HALF_ST_DM3   =: ');
IN_DM3      =1;%input('encoder input      IN_DM3   =: ');
N_out_DM3   =2;%input('encoder output      N_out_DM3 =: ');
K_DM3       =7;%input('the constraint len.      K_DM3 =: ');
L_DM3       =750;%input('data length              L_DM3 =: ');
M_DM3       =6;%input('TAIL length              L_DM3 =: ');
%#################################################################
%####################     DM5  CONV        #######################
STATE_N_DM5 =64;%input('STATE NO        STATE_N_DM5   =: ');
HALF_ST_DM5 =32;%input('1/2STATE NO        HALF_ST_DM5   =: ');
IN_DM5      =1;%input('encoder input      IN_DM5   =: ');
N_out_DM5   =2;%input('encoder output      N_out_DM5 =: ');
K_DM5       =7;%input('the constraint len.      K_DM5 =: ');
L_DM5       =1372;%input('data length              L_DM5 =: ');
M_DM5       =6;%input('TAIL length              L_DM5 =: ');
%#################################################################
%####################     PROGRAM VARIABLE
##################

n2=63    ;% input('the no of codeword bch-access  =  ');
%n2=64
k2=30    ;% input('the no of k data bch encoder =  ');        %
k2=30
n1=15;%input('the no of code word DM  n1=  ');
k1=11;%input('the no of data word DM  k1=  ');
N=input('the no of iteration         N=  ');
%#################################################################
%#################################################################

%G IS GENERATOR MATRIX FOR HAMMING (15,10)ENCODER
%G=[1 1 0 1 0 1 0 0 0 0 0 0 0 0 0;
```

```
        %0 1 1 0 1 0 1 0 0 0 0 0 0 0 0;
        %1 1 1 0 0 0 0 1 0 0 0 0 0 0 0;
        %0 1 1 1 0 0 0 0 1 0 0 0 0 0 0;
        %0 0 1 1 1 0 0 0 0 1 0 0 0 0 0;
        %1 1 0 0 1 0 0 0 0 0 1 0 0 0 0;
        %1 0 1 1 0 0 0 0 0 0 0 1 0 0 0;
        %0 1 0 1 1 0 0 0 0 0 0 0 1 0 0;
        %1 1 1 1 1 0 0 0 0 0 0 0 0 1 0;
        %1 0 1 0 1 0 0 0 0 0 0 0 0 0 1];

      % H  IS PARITY_CHECK MATRIX FOR HAMMING (15,10) DECODER

      %H=[1 0 0 0 0 1 0 1 0 0 1 1 0 1 1;
         % 0 1 0 0 0 1 1 1 1 0 1 0 1 1 0;
          %0 0 1 0 0 0 1 1 1 1 0 1 0 1 1;
          %0 0 0 1 0 1 0 0 1 1 0 1 1 1 0;
          %0 0 0 0 1 0 1 0 0 1 1 0 1 1 1];
    %#################################################################
    %#################################################################
    %#################################################################

    counter_itr=0;
    snr=[1:1:6];
    y=length(snr);
    snrabs=[];
    for j=1:y

        snrabs(j)=10^(snr(j)/10);
        sigma=1/sqrt(2*snrabs(j));
        error_dh1=0;
        error_dm1=0;
        error_dh3=0;
        error_dm3=0;
        error_dh5=0;
        error_dm5=0;
        per1=0;
        per2=0;
        per3=0;
        counter_itr=counter_itr+1
        for i=1:N

    %p1=round(rand(1,4));                %preample
    %p2=round(rand(1,30));                %sync word LAP=24+6
    %p3=round(rand(1,4));                %trailer
    y_AC=round(rand(1,L_AC));
        for x=1:M_AC
      y_AC(L_AC+M_AC)=0;
        end
    %p4=round(rand(1,18));                %header
     y_HD=round(rand(1,L_HD));
        for x=1:M_HD
      y_HD(L_HD+M_HD)=0;
        end

    %dm1=round(rand(1,176));              %payload dm1 packet fec
(15,10)ham
```

```
    y_DM1=round(rand(1,L_DM1));
        for x=1:M_DM1
      y_DM1(L_DM1+M_DM1)=0;
        end
    %dm3=round(rand(1,1001));              %payload dm3 packet ,,,,,,,,,,,
    y_DM3=round(rand(1,L_DM3));
        for x=1:M_DM3
      y_DM3(L_DM3+M_DM3)=0;
        end
    %dm5=round(rand(1,2013));              %payload dm5 packet,,,,,,,,,,,
    y_DM5=round(rand(1,L_DM5));
        for x=1:M_DM5
      y_DM5(L_DM5+M_DM5)=0;
        end
    %##################################################################
    %bt_pkt1=[p2,p4,dm1];
    %bt_pkt3=[p2,p4,dm3];
    %bt_pkt5=[p2,p4,dm5];
    %######################### WITH CONV IN ENCO #######################
    bt_pkt1=[y_AC,y_HD,y_DM1];
    bt_pkt3=[y_AC,y_HD,y_DM3];
    bt_pkt5=[y_AC,y_HD,y_DM5];

    %-----------------------------------=


    %bt_pkth1=[p1,p2,p3,p4,dh1];
    %bt_pkth3=[p1,p2,p3,p4,dh3];
    %bt_pkth5=[p1,p2,p3,p4,dh5];
    %****************************************************************
    % p2 sync word use code bch code(64,30)

     %p2=gf(p2);              %BCH CODE ACCESS CODE PART IN GENERAL FORM
     %[genpoly,t]=bchgenpoly(n2,k2);
     %pg = bchpoly(n2, k2);
     %p2_encode=bchenco(p2,n2,k2);
                %enc_word_AC=p2_encode
                %#########  CONV  ENCODER

[enc_word_AC,sym_AC,data_len_AC]=conv_encode_AC(STATE_N_AC,M_AC,IN_AC,
N_out_AC,K_AC,L_AC,y_AC);
    % P4 IS HEADER USE REPETITION CODE 1/3
    %for k=1:18
     %   for l=1:3
      %       p4_encode(3*(k-1)+1)=p4(k);
      %end
      %end
                %enc_word_HD=p4_encode
                %#########  CONV  ENCODER

[enc_word_HD,sym_HD,data_len_HD]=conv_encode_HD(STATE_N_HD,M_HD,IN_HD,
N_out_HD,K_HD,L_HD,y_HD);

    %********************************************
    %P5=PDM OR PDH OR OTHER DM USE HAMM.CODE (15,10)

    %dm1_encode=encode(dm1,n1,k1,'hamming/binary');
```

```
%ENCODING(X,G,SEGM) IT IS FUNCTION
    %dm3_encode=encode(dm3,n1,k1,'hamming/binary');
    %dm5_encode=encode(dm5,n1,k1,'hamming/binary');
                     %#########  CONV  ENCODER

[enc_word_DM1,sym_DM1,data_len_DM1]=conv_encode_DM1(STATE_N_DM1,M_DM1,
IN_DM1,N_out_DM1,K_DM1,L_DM1,y_DM1);


[enc_word_DM3,sym_DM3,data_len_DM3]=conv_encode_DM3(STATE_N_DM3,M_DM3,
IN_DM3,N_out_DM3,K_DM3,L_DM3,y_DM3);


[enc_word_DM5,sym_DM5,data_len_DM5]=conv_encode_DM5(STATE_N_DM5,M_DM5,
IN_DM5,N_out_DM5,K_DM5,L_DM5,y_DM5);

    %*****************************************************
    %so the bt_pkt after encoding & before sending  it as will be
    %bt_pkt_enc1=[p2_encode,p4_encode,dm1_encode1];
    %bt_pkt_enc3=[p2_encode,p4_encode,dm1_encode3];
    %bt_pkt_enc5=[p2_encode,p4_encode,dm1_encode5];
    %------------------------------------
     %#########  CONV  ENCODER
     bt_pkt_enc1=[enc_word_AC,enc_word_HD,enc_word_DM1];
     bt_pkt_enc3=[enc_word_AC,enc_word_HD,enc_word_DM3];
     bt_pkt_enc5=[enc_word_AC,enc_word_HD,enc_word_DM5];

    %              MOD. AS BPSK FOR ALL PACKET TYPE
    %                      MODULATION

    mod_bt_pkt1=2*bt_pkt_enc1-1;
    mod_bt_pkt3=2*bt_pkt_enc3-1;
    mod_bt_pkt5=2*bt_pkt_enc5-1;


%######################################################################

    %                   CHANNEL EFFECTS
    %##################   AWGN CHANNEL
##################################
    rec_bt_pkt1=mod_bt_pkt1+sigma*randn(1,length(bt_pkt_enc1));
    rec_bt_pkt3=mod_bt_pkt3+sigma*randn(1,length(bt_pkt_enc3));
    rec_bt_pkt5=mod_bt_pkt5+sigma*randn(1,length(bt_pkt_enc5));

    %###################### SLOW FADING
#################################
   % deep_fad=sqrt((0.5*(rand(1,1)^2))+(0.5*(randn(1,1)^2)));
   % ##############
   % bt_pkt_d_fad1=mod_bt_pkt1*deep_fad;
    %rec_bt_pkt1=bt_pkt_d_fad1+sigma*randn(1,length(bt_pkt_enc1));
    %##################
    %bt_pkt_d_fad3=mod_bt_pkt3*deep_fad;
   % rec_bt_pkt3=bt_pkt_d_fad3+sigma*randn(1,length(bt_pkt_enc3));
    %###################
```

```
    % bt_pkt_d_fad5=mod_bt_pkt5*deep_fad;
    % rec_bt_pkt5=bt_pkt_d_fad5+sigma*randn(1,length(bt_pkt_enc5));


    %                          DEMODULATION

    % demod_bt_pkt1=rec_bt_pkt1>0;
    %________________________
     for r=1:length(rec_bt_pkt1)
         if rec_bt_pkt1(r)>0

    demod_bt_pkt1(r)=1;
else
    demod_bt_pkt1(r)=0;
end
end
    %demod_bt_pkt3=rec_bt_pkt3>0;
    %________________________________
     for r=1:length(rec_bt_pkt3)
         if rec_bt_pkt3(r)>0

    demod_bt_pkt3(r)=1;
else
    demod_bt_pkt3(r)=0;
end
end
   % demod_bt_pkt5=rec_bt_pkt5>0;
   %___________________________________
    for r=1:length(rec_bt_pkt5)
        if rec_bt_pkt5(r)>0

    demod_bt_pkt5(r)=1;
else
    demod_bt_pkt5(r)=0;
end
end


    %                              DECODING


    %                                  DM1

    %p1_rec1=demod_bt_pkt1(1,1:4);
    p2_rec1=demod_bt_pkt1(1,1:length(enc_word_AC));     %IT FOR DH
PKT ONE FROM DM  ONE FROM DH
    %p3_rec1=demod_bt_pkt1(1,68:71);

p4_rec1=demod_bt_pkt1(1,1+length(enc_word_AC):length(enc_word_AC)+leng
th(enc_word_HD));

p5_rec1=demod_bt_pkt1(1,1+length(enc_word_AC)+length(enc_word_HD):leng
th(bt_pkt_enc1));
```

```
%##########################)

 %p1_dec1=p1_rec1;
 %p2_dec1=bchdeco(p2_rec1,k2,6);
 %p3_dec1=p3_rec1;
% ################CONV DECODER

con_bt_AC1=conv_decode_AC1(HALF_ST_AC,data_len_AC,N_out_AC,K_AC,p2_rec
1,sym_AC,y_AC);
   p2_dec1=con_bt_AC1 ;  %############------------->

 %p4 IS HEADER REPETION CODE
%_________________________________________
%for k=1:18
 %   counter=0;
  %  for l=1:3
   %     if p4_rec1(3*(k-1)+l)==1
    %        counter=counter+1;
    %  end
    %end
    %if counter >1
     %    p4_dec1(k)=1;
     %else
     %    p4_dec1(k)=0;
     %end
     % end
       % ################CONV DECODER

con_bt_HD1=conv_decode_HD1(HALF_ST_HD,data_len_HD,N_out_HD,K_HD,p4_rec
1,sym_HD,y_HD);
                p4_dec1=con_bt_HD1;    %############-------------->
    %##########################
      %p5_dec1=decode(p5_rec1',n1,k1,'hamming/binary');
    p5_rec1=enc_word_DM1;

con_bt_DM1=conv_decode_DM1(HALF_ST_DM1,data_len_DM1,N_out_DM1,K_DM1,en
c_word_DM1,sym_DM1,y_DM1);

    %                  DM3

%p1_rec3=demod_bt_pkt3(1,1:4);
p2_rec3=demod_bt_pkt3(1,1:length(enc_word_AC));       %IT FOR DH
PKT ONE FROM DM  ONE FROM DH
%p3_rec3=demod_bt_pkt3(1,68:71);

p4_rec3=demod_bt_pkt3(1,1+length(enc_word_AC):length(enc_word_AC)+leng
th(enc_word_HD));

p5_rec3=demod_bt_pkt3(1,1+length(enc_word_AC)+length(enc_word_HD):leng
th(bt_pkt_enc3));

%###########################

%p1_dec3=p1_rec3;
%p2_dec3=bchdeco(p2_rec3,k2,6);
%p3_dec3=p3_rec3;
% ################CONV DECODER
```

```
con_bt_AC3=conv_decode_AC3(HALF_ST_AC,data_len_AC,N_out_AC,K_AC,p2_rec
3,sym_AC,y_AC);
      p2_dec3=con_bt_AC3;            %#################-------------->
   %_____________________________

    %p4 IS HEADER REPETION CODE
   %________________________________________
   %for k=1:18
    %   counter=0;
    %
     %    for l=1:3
      %      if p4_rec3(3*(k-1)+1)==1
       %         counter=counter+1;
       %  end
       %end
       %if counter >1
        %    p4_dec3(k)=1;
        %else
        %    p4_dec3(k)=0;
        %end
        %end
      % ################CONV DECODER

con_bt_HD3=conv_decode_HD3(HALF_ST_HD,data_len HD,N_out_HD,K_HD,p4_rec
3,sym_HD,y_HD);
              p4_dec3=con_bt_HD3 ;  %############-------------->
      %##########################
         %p5_dec3=decode(p5_rec3',n1,k1,'hamming/binary');
      enc_word_DM3=p5_rec3;

con_bt_DM3=conv_decode_DM3(HALF_ST_DM3,data_len_DM3,K_DM3,N_out_DM3,en
c_word_DM3,sym_DM3,y_DM3);

      %                           DM5

   %p1_rec5=demod_bt_pkt5(1,1:4);
   p2_rec5=demod_bt_pkt5(1,1:length(enc_word_AC));       %IT FOR DH
PKT ONE FROM DM  ONE FROM DH
   %p3_rec5=demod_bt_pkt5(1,68:71);

p4_rec5=demod_bt_pkt5(1,1+length(enc_word_AC):length(enc_word_AC)+leng
th(enc_word_HD));

p5_rec5=demod_bt_pkt5(1,1+length(enc_word_AC)+length(enc_word_HD):leng
th(bt_pkt_enc5));

      %##########################

      %p1_dec5=p1_rec5;
      %p2_dec5=bchdeco(p2_rec5,k2,6);
      %p3_dec5=p3_rec5;
       % ################CONV DECODER
```

```
con_bt_AC5=conv_decode_AC5(HALF_ST_AC,data_len_AC,N_out_AC,K_AC,p2_rec
5,sym_AC,y_AC);
    p2_dec5=con_bt_AC5;          %#################------------->
    %___________________________

     %p4 IS HEADER REPETION CODE
    %______________________________________

    %for k=1:18
      % counter=0;
        %for l=1:3
            %if p4_rec5(3*(k-1)+l)==1
                %counter=counter+1;
                % end
           %end
        %if counter >1
           % p4_dec5(k)=1;
            %else
            %p4_dec5(k)=0;
            %end
         %end
      % ################CONV DECODER

con_bt_HD5=conv_decode_HD5(HALF_ST_HD,data_len_HD,N_out_HD,K_HD,p4_rec
5,sym_HD,y_HD);
                p4_dec5=con_bt_HD5;    %############-------------->
     %###########################
         % p5_dec5=decode(p5_rec5',n1,k1,'hamming/binary');
         enc_word_DM5=p5_rec5;

con_bt_DM5=conv_decode_DM5(HALF_ST_DM5,data_len_DM5,K_DM5,N_out_DM5,en
c_word_DM5,sym_DM5,y_DM5);
 %###############################################################
 %@###############################################################
         bt_pkt_dec1=[p2_dec1,p4_dec1,con_bt_DM1];
         bt_pkt_dec3=[p2_dec3,p4_dec3,con_bt_DM3];
         bt_pkt_dec5=[p2_dec5,p4_dec5,con_bt_DM5];
   %#################################################################

       x1=0;
       x2=0;
       x3=0;
         x1=sum(xor(bt_pkt_dec1, bt_pkt1));
         error_dm1=error_dm1+x1;
         x2=sum(xor(bt_pkt_dec3, bt_pkt3));
         error_dm3=error_dm3+x2;
         x3=sum(xor(bt_pkt_dec5, bt_pkt5));
         error_dm5=error_dm5+x3;

         if x1>0
             per1=per1+1;
         end
              if x2>0
             per2=per2+1;
         end
          if x3>0
             per3=per3+1;
         end
```

```
%##############################################################

%##############################################################

    end

%##############################################################

%##############################################################
    .   ber_dm1(j)=error_dm1/(length(bt_pkt1)*N);
        ber_dm3(j)=error_dm3/(length(bt_pkt3)*N); %NO OF ITERATION*NO
OF BITS IN PKT
        ber_dm5(j)=error_dm5/(length(bt_pkt5)*N); %NO OF ITERATION*NO
OF BITS IN PKT
        per_dm1(j)=per1/N;
        per_dm3(j)=per2/N;
        per_dm5(j)=per3/N;
    end
      figure(1);
      hold on
       semilogy(snr,ber_dm1,'s-k');
       semilogy(snr,ber_dm3,'^-r');
       semilogy(snr,ber_dm5,'b');
       legend('DM1 Packets','DM3 Packets','DM5 Packets');
       title('SNR  TO  BER  AWGN DMx with CONV=1/2 AND K=87]');
       xlabel('Eb/No (dB)');
       ylabel('Bit Error Rate');
       grid on;
       figure(2);
       semilogy(snr,per_dm1,'s-k');
       hold on
       semilogy(snr,per_dm3,'^-r');
       hold on
       semilogy(snr,per_dm5);
       legend('DM1 Packets','DM3 Packets','DM5 Packets');
       title('SNR  TO  PER  AWGN DMx with CONV=1/2 AND K=7]');
       xlabel('Eb/No (dB)');
       ylabel('Packet Error Rate');
       grid on
%##############################################################
```

**The Required Functions**:

***The Shift Function**:

```
%##################################################
function state=shift(v,w,state)
for a=w+1:-1;2
    state(a)=state(a-1);
end
state(1)=v;
      %fun.
%##################################################
```

***The encoding function code**:

```
%###################################################
  %    INPUTS DEFINITIONS

%IN_DM5       =input('encoder input         IN_DM5   =: ');
%N_out_DM5    =input('encoder output        N_out_DM5 =: ');
%K_DM5        =input('the constraint len.       K_DM5 =: ');
%L_DM5        =input('data length               L_DM5 =: ');


function
[enc_word_DM5,sym_DM5,data_len_DM5]=conv_encode_DM5(STATE_N_DM5,M_DM5,
IN_DM5,N_out_DM5,K_DM5,L_DM5,y_DM5);


switch N_out_DM5

    case {2}

        switch K_DM5

            case {3}
                gen_fun=[1 0 1;1 1 1];
            case {4}
                gen_fun=[1 1 0 1;1 1 1 1];
            case {5}
                gen_fun=[1 0 0 1 1;1 1 1 0 1];
            case {6}
                gen_fun=[1 1 0 1 0 1;1 0 1 1 1 1];
            case {7}
                gen_fun=[1 0 1 1 0 1 1;1 1 1 1 0 0 1];
            case {8}
                gen_fun=[1 1 1 0 0 1 0 1;1 0 0 1 1 1 1 1 ];
            case {9}
                gen_fun=[1 0 1 1 1 0 0 0 1;1 1 1 1 0 1 0 1 1];
            case {10}
                gen_fun=[1 0 0 1 1 1 0 1 1 1;1 1 0 1 1 0 0 1 0 1];
            case {11}
                gen_fun=[1 0 0 1 1 0 1 1 1 0 1;1 1 1 1 0 1 1 0 0 0 1];
            case {12}
                gen_fun=[1 0 0 0 1 1 0 1 1 1 0 1;1 0 1 1 1 1 0 1 0 0 1
1];
            case {13}
                gen_fun=[1 0 0 0 1 0 1 0 1 1 0 1 1;1 1 1 1 1 1 0 1 1 0
0 0 1];
            case {14}
                gen_fun=[1 0 0 0 1 1 1 0 1 1 1 1 0 1;1 0 1 1 1 0 0 1 0
1 0 0 1 1];
            case {15}
                gen_fun=[1 0 1 1 1 0 1 1 1 0 1 0 0 0 1;1 1 0 0 0 1 1 1
1 0 0 1 0 1 1];
            case {16}
                gen_fun=[1 0 0 1 0 0 1 1 1 0 1 0 1 0 1 1;1 1 0 0 1 0 1
1 1 0 1 1 0 1 0 1];
            case {17}
                gen_fun=[1 1 1 0 0 1 1 1 0 1 0 1 0 0 0 0 1;1 0 1 0 0 1
```

```
1 0 0 1 0 1 1 1 1 1 1];


          otherwise

           disp('THE VALUE HIGHER THAN-17- OR LESS THAN -3- IT IS
OUT OF RANGE');


    end             %======== END R=1\2 WITH K= 3---8


 case {3}           %========CASE [3] Nout = 3

        switch K_DM5

          case {3}
              gen_fun=[1 0 1;1 1 1;1 1 1];
          case {4}
              gen_fun=[1 1 0 1;1 1 0 1;1 1 1 1];
          case {5}
              gen_fun=[1 0 1 0 1;1 1 0 1 1;1 1 1 1 1];
          case {6}
              gen_fun=[1 0 0 1 1 1;1 0 1 0 1 1;1 1 1 1 0 1];
          case {7}
              gen_fun=[1 0 1 1 0 1 1;1 1 0 0 1 0 1;1 1 1 1 1 0 1];
          case {8}
              gen_fun=[1 0 0 1 0 1 0 1;1 1 0 1 1 0 0 1;1 1 1 1 0 1 1
1 ];
            case {9}
              gen_fun=[1 0 1 1 0 1 1 1 1;1 1 0 1 1 0 0 1 1;1 1 1 0 0
1 0 0 1];
          case {10}
              gen_fun=[1 0 0 1 0 0 1 1 1 1;1 0 1 1 1 1 0 1 0 1;1 1 1
0 0 1 1 0 1 1];
          case {11}
              gen_fun=[1 0 0 1 1 1 0 1 0 1 1;1 0 1 1 0 1 1 1 0 0 1;1
1 0 0 1 1 1 1 1 0 1];
          case {12}
              gen_fun=[1 0 0 1 1 1 1 1 0 1 1 1;1 0 1 1 1 1 0 1 0 0 1
1
                      1 1 0 0 1 0 1 1 0 1 0 1];
          case {13}
              gen_fun=[1 0 0 0 1 0 1 0 1 1 0 1 1;1 0 0 0 1 1 0 1 1 1
1 0 1
                      1 1 1 1 1 1 0 1 1 0 0 0 1];
          case {14}
              gen_fun=[1 0 0 0 1 1 1 0 1 0 0 1 0 1;1 1 1 0 1 1 1 0 1
1 0 0 0 1
                      1 1 1 1 1 0 0 1 0 1 1 0 1 1];

          otherwise

              disp('THE VALUE HIGHER THAN-14- OR LESS THAN -3- IT IS
```

```
OUT OF RANGE {3--14');

        end

      case {4}

          switch K_DM5

              case {3}
                   gen_fun=[1 0 1;1 1 1;1 1 1;1 1 1];
              case {4}
                  gen_fun=[1 0 1 1;1 1 0 1;1 1 0 1;1 1 1 1];
              case {5}
                  gen_fun=[1 0 1 0 1;1 0 1 1 1;1 1 0 1 1;1 1 1 1 1];
              case {6}
                  gen_fun=[1 0 1 0 1 1;1 1 0 1 1 1;1 1 1 0 0 1;1 1 1 1
0 1];
              case{7}
                  gen_fun=[1 0 1 1 1 0 1;1 0 1 1 1 0 1;1 1 0 0 1 1 1;1
1 1 0 0 1 1];
              case {8}
                  gen_fun=[1 0 0 1 1 1 0 1;1 0 1 1 1 1 0 1;1 1 0 0 1 0
1 1;1 1 1 0 1 1 1 1];
              case {9}
                gen_fun=[1 0 0 1 1 0 0 1 1;1 0 1 0 1 1 1 0 1
                         1 1 1 0 1 1 0 1 1;1 1 1 1 0 0 1 0 1];
              case {10}
                gen_fun=[1 0 0 1 0 0 1 1 1 1;1 0 1 1 1 1 0 1 0 1
                         1 1 1 0 0 1 1 0 1 1;1 1 1 0 1 0 1 0 1 1];
              case {11}
                gen_fun=[1 0 0 1 1 0 1 0 1 1 1;1 0 0 1 1 1 0 1 0 1 1
                         1 0 1 1 0 1 1 1 0 0 1;1 1 0 0 1 1 1 1 1 0 1];
              case {12}
                gen_fun=[1 0 0 1 1 1 1 1 0 1 1 1;1 0 1 1 1 1 0 1 0 0 1
1
                         1 1 0 0 1 0 1 1 0 1 0 1;1 1 1 1 0 0 1 0 1 1 0
1];
              case {13}
                gen_fun=[1 0 0 1 0 0 1 1 0 0 1 0 1;1 0 1 0 1 0 0 1 1 1
1 1 1
                         1 1 0 1 1 0 1 1 1 1 0 1 1;1 1 1 0 1 1 1 0 1 0
1 1 1];
              case {14}
                gen_fun=[1 0 0 0 1 0 0 1 0 0 1 0 1 1;1 0 0 1 1 0 0 1 1
1 1 1 0 1
                         1 1 1 0 1 1 0 1 0 1 0 1 1 1;1 1 1 0 1 1 0 1 0
1 1 1 1 1];

              otherwise

                  disp('THE VALUE HIGHER THAN-14- OR LESS THAN -3- IT
IS OUT OF RANGE {3--14');
      end

end
```

```
%        STARTING VALUE

%======================================================================


state_tab=zeros(2*STATE_N_DM5,N_out_DM5+1);
state    =[];
mq       =1;

%t_state =[0 0 0;0 1 0;0 0 1;0 1 1];
% T_STATE IT IS HERE TO CHECK PROGRAM WHERE {R=1\2----K=3} }

%        STATE TABEL PART
%========================


for a=1:STATE_N_DM5
      a=a-1;
    for i=2:K_DM5
        state(1)=0;
        state(i)=rem(a,2);
        if a>0
            a=round(a/2-0.5);
        else
            a=0;

        end

        %  disp(['the state form as in enoder
process=:',num2str(state)]);

    end

    %  state=t_state(a,:);

    for b=0:IN_DM5
    state=shift(b,M_DM5,state);

    for c=1:N_out_DM5

        for d=1:K_DM5

state_tab(mq,c)=xor(state_tab(mq,c),(gen_fun(c,d)&state(d)));
        end                             % d loop

    end                                 % c loop

            c=1;                        %NEXT STATE CACULATE
              for d=1:M_DM5
                c=c+state(d)*(2^(d-1));
            end
                state_tab(mq,N_out_DM5+1)=c;
                c=1;
```

```
            for  d=1:N_out_DM5                  %SYM. INDEX DENOTE TO
METS INDEX
                c=c+(2^(d-1))*state_tab(mq,d);        %IT MAY IN
STAT_TAB AS
                sym_DM5(mq)=c;
%COLUMN [U+1]
            end
            mq=mq+1;
        end             %  loop b

    end                 %  loop a


    %
=======================================================================
=

    %   CHECK POINT --- A
%disp(['the state table of this case, stat_tab(mq,c)=:
',num2str(stat_tab)]);

%
=======================================================================

 %EbN0=[0:6];
%loop=1000;
%for SNR=1:length(EbN0)
 %    SIGMA=sqrt(N_out/2*EbN0(SNR));
  %   for i=1:loop



  %        DATA GENERATION & ADD TAIL M=ZEROES

  %      ======================================

  %y=round(rand(1,L));
 %for x=1:M
  %     y(L+M)=0;
  %end

 %
=======================================================================
===

 %    CHECK POINT ---- B


 %disp(['the data series y=L+m ,y=: ',num2str(y)]);
 %
=======================================================================
==
```

```
%           ENCODING PART
%         ======================


s=1;
data_len_DM5=length(y_DM5);

for i=1:data_len_DM5                       %leng=L+m

    switch N_out_DM5

        case{2}
            enc_word_DM5(2*i-1)=state_tab(2*s-1+y_DM5(i),1);
            enc_word_DM5(2*i)  =state_tab(2*s-1+y_DM5(i),2);
            s                  =state_tab(2*s-1+y_DM5(i),3);
        case{3}
            enc_word_DM5(3*i-2)=state_tab(2*s-1+y_DM5(i),1);
            enc_word_DM5(3*i-1)=state_tab(2*s-1+y_DM5(i),2);
            enc_word_DM5(3*i)  =state_tab(2*s-1+y_DM5(i),3);
            s                  =state_tab(2*s-1+y_DM5(i),4);
        case{4}
            enc_word_DM5(4*i-3)=state_tab(2*s-1+y_DM5(i),1);
            enc_word_DM5(4*i-2)=state_tab(2*s-1+y_DM5(i),2);
            enc_word_DM5(4*i-1)=state_tab(2*s-1+y_DM5(i),3);
            enc_word_DM5(4*i)  =state_tab(2*s-1+y_DM5(i),4);
            s                  =state_tab(2*s-1+y_DM5(i),5);

        otherwise
            disp('FOR N_out>3 THERE IS OTHER RULE ');

    end

end
%###################################################
```

```
%####################################################
```

**The decoding Function of the decoding**:

```
%#################################################
         %    INPUTS DEFINITIONS
%IN_DM5      =input('encoder input       IN_DM5  =: ');
%N_out_DM5   =input('encoder output      N_out_DM5 =: ');
%K_DM5       =input('the constraint len.     K_DM5 =: ');
%L_DM5       =input('data length             L_DM5 =: ');

function
con_bt_DM5=conv_decode_DM5(HALF_ST_DM5,data_len_DM5,K_DM5,N_out_DM5,en
c_word_DM5,sym_DM5,y_DM5)



 %              DECODING PART
 %          =========================


 met_tab=[0 1;1 0];
 statem =[0 5 5 5 5 --------------------5 up to more than the longer
encoded data length];

   for x=1:data_len_DM5
      %y=L+m data encoder input length
          mets=zeros(2^N_out_DM5);

          switch N_out_DM5

              case{2}
              sub_code(1)=enc_word_DM5(2*x-1);
              sub_code(2)=enc_word_DM5(2*x);
              mets=zeros(2^N_out_DM5);
            mets(1)=met_tab(1,1+sub_code(1)) +
met_tab(1,1+sub_code(2));
            mets(2)=met_tab(2,1+sub_code(1)) +
met_tab(1,1+sub_code(2));
            mets(3)=met_tab(1,1+sub_code(1)) +
met_tab(2,1+sub_code(2));
            mets(4)=met_tab(2,1+sub_code(1)) +
met_tab(2,1+sub_code(2));

        case{3}
            sub_code(1)=enc_word_DM5(3*x-2);
            sub_code(2)=enc_word_DM5(3*x-1);
            sub_code(3)=enc_word_DM5(3*x);
              mets=zeros(2^N_out_DM5);
        mets(1)=met_tab(1,1+sub_code(1)) +
met_tab(1,1+sub_code(2))+met_tab(1,1+sub_code(3));
        mets(2)=met_tab(2,1+sub_code(1)) +
met_tab(1,1+sub_code(2))+met_tab(1,1+sub_code(3));
        mets(3)=met_tab(1,1+sub_code(1)) +
met_tab(2,1+sub_code(2))+met_tab(1,1+sub_code(3));
        mets(4)=met_tab(2,1+sub_code(1)) +
met_tab(2,1+sub_code(2))+met_tab(1,1+sub_code(3));
        mets(5)=met_tab(1,1+sub_code(1)) +
```

```
met_tab(1,1+sub_code(2))+met_tab(2,1+sub_code(3));
        mets(6)=met_tab(2,1+sub_code(1)) +
met_tab(1,1+sub_code(2))+met_tab(2,1+sub_code(3));
        mets(7)=met_tab(1,1+sub_code(1)) +
met_tab(2,1+sub_code(2))+met_tab(2,1+sub_code(3));
        mets(8)=met_tab(2,1+sub_code(1)) +
met_tab(2,1+sub_code(2))+met_tab(2,1+sub_code(3));

       case{4}
           sub_code(1)=enc_word_DM5(4*x-3);
           sub_code(2)=enc_word_DM5(4*x-2);
           sub_code(3)=enc_word_DM5(4*x-1);
           sub_code(4)=enc_word_DM5(4*x);
             mets=zeros(2^N_out_DM5);
mets(1) =met_tab(1,1+sub_code(1)) + met_tab(1,1+sub_code(2))
+met_tab(1,1+sub_code(3)) +met_tab(1,1+sub_code(4));
mets(2) =met_tab(2,1+sub_code(1)) + met_tab(1,1+sub_code(2))
+met_tab(1,1+sub_code(3)) +met_tab(1,1+sub_code(4));
mets(3) =met_tab(1,1+sub_code(1)) + met_tab(2,1+sub_code(2))
+met_tab(1,1+sub_code(3)) +met_tab(1,1+sub_code(4));
mets(4) =met_tab(2,1+sub_code(1)) + met_tab(2,1+sub_code(2))
+met_tab(1,1+sub_code(3)) +met_tab(1,1+sub_code(4));
mets(5) =met_tab(1,1+sub_code(1)) + met_tab(1,1+sub_code(2))
+met_tab(2,1+sub_code(3)) +met_tab(1,1+sub_code(4));
mets(6) =met_tab(2,1+sub_code(1)) + met_tab(1,1+sub_code(2))
+met_tab(2,1+sub_code(3)) +met_tab(1,1+sub_code(4));
mets(7) =met_tab(1,1+sub_code(1)) + met_tab(2,1+sub_code(2))
+met_tab(2,1+sub_code(3)) +met_tab(1,1+sub_code(4));
mets(8) =met_tab(2,1+sub_code(1)) + met_tab(2,1+sub_code(2))
+met_tab(2,1+sub_code(3)) +met_tab(1,1+sub_code(4));
mets(9) =met_tab(1,1+sub_code(1)) + met_tab(1,1+sub_code(2))
+met_tab(1,1+sub_code(3)) +met_tab(2,1+sub_code(4));
mets(10)=met_tab(2,1+sub_code(1)) + met_tab(1,1+sub_code(2))
+met_tab(1,1+sub_code(3)) +met_tab(2,1+sub_code(4));
mets(11)=met_tab(1,1+sub_code(1)) + met_tab(2,1+sub_code(2))
+met_tab(1,1+sub_code(3)) +met_tab(2,1+sub_code(4));
mets(12)=met_tab(2,1+sub_code(1)) + met_tab(2,1+sub_code(2))
+met_tab(1,1+sub_code(3)) +met_tab(2,1+sub_code(4));
mets(13)=met_tab(1,1+sub_code(1)) + met_tab(1,1+sub_code(2))
+met_tab(2,1+sub_code(3)) +met_tab(2,1+sub_code(4));
mets(14)=met_tab(2,1+sub_code(1)) + met_tab(1,1+sub_code(2))
+met_tab(2,1+sub_code(3)) +met_tab(2,1+sub_code(4));
mets(15)=met_tab(1,1+sub_code(1)) + met_tab(2,1+sub_code(2))
+met_tab(2,1+sub_code(3)) +met_tab(2,1+sub_code(4));
mets(16)=met_tab(2,1+sub_code(1)) + met_tab(2,1+sub_code(2))
+met_tab(2,1+sub_code(3)) +met_tab(2,1+sub_code(4));

otherwise

    disp('FOR N_out>4 THERE IS SOME OTHER RULE');

end

for i=1:HALF_ST_DM5
```

```
                m0=statem(i) + mets(sym_DM5(2*i-1));
                m1=statem(i+HALF_ST_DM5) +
mets(sym_DM5(2*(i+HALF_ST_DM5)-1));


                if m0<=m1
                    nextm(2*i-1)      =m0;
                    prev_tab(2*i-1,x)=i;
                else
                    nextm(2*i-1)      =m1;
                    prev_tab(2*i-1,x)=i+HALF_ST_DM5;

                end

                m0=statem(i) + mets(sym_DM5(2*i));
                m1=statem(i+HALF_ST_DM5) +
mets(sym_DM5(2*(i+HALF_ST_DM5)));

                if m0<=m1
                    nextm(2*i)      =m0;
                    prev_tab(2*i,x)=i;
                else
                    nextm(2*i)      =m1;
                    prev_tab(2*i,x)=i+HALF_ST_DM5;

                end

            end

            statem=nextm;

            % disp(['the new present state is statem=:
',num2str(statem)])

        end

        %
=================================================================

        % THE LAST CHECK POINT AFTER TRACE BACK
      %  CHECK POINT ---- D


      %    disp(['the prev_tab=: ',prev_tab]);

      %fprintf(prev_tab,'%6.2f');


  %  =================================================================
```

```
% TRACE BACK PART
% ====================
%trace back

best_state=1;
for a=data_len_DM5:-1:1
    con_bt_DM5(a)=mod(best_state-1,2);
    best_state=prev_tab(best_state,a);

end

%disp(['the output of decoder ,con_bt(a)=: ',num2str(con_bt)]);


% FINAL CHECK
%COMPARE BETWEEN THE ENCODER INPUT(Y=L+m)is data_in AND THE DECODER
%OUTPUT WHICH IS decodxe(a)


%
disp('=====================================================================
=========');
%disp(['the data series y=L+m         ,y=: ',num2str(y)]);
%disp(['the output of decoder ,decode(a)=: ',num2str(decode)]);
%disp('================================
%###########################################################
```

The uncoded packet simulation codes for the modified packet are presented as follows:

The standard uncoded packets simulation codes are the same as presented in the previous chapter.

%%####################################################
#######

## 5.7 Complexity of Convolutional Codes

The convolutional codes are widely used in the wireless communications systems. It is the main content of the Turbo code. Also, it is considered that complex error control scheme, the processed data, and the packet length are important factors in the computational complexity equation [13–16].

In [13], the complexity of convolutional codes is discussed and it cleared the amount of complexity due to the processed data and the memory depth ones. Also, the analysis focuses on the length of constraint length which plays role in the convolutional codes performance and the complexity degree.

The input information sequence of the convolutional codes contains $k \times L$ bits, where k is the number of parallel information bits at a time interval, and L is the number of time intervals. This results in $m + L$ stages in the trellis diagram, where $m$ is the number of shift registers in the encoder and $K = m + 1$. There are exactly

$2^{k \times L}$ distinct paths in the trellis diagram. As a result, the maximum likelihood (ML) sequence would have a computational complexity of O ($2^{k \times L}$). The Viterbi algorithm reduces complexity by performing the ML search on a stage at a time in the trellis diagram at each node [15]. The number of nodes per stage in the trellis diagram is $2^m$. Therefore, the complexity of Viterbi calculations is of O (($2^k$) ($2^m$) ($m + L$)). This significantly reduces the number of calculations required to implement the ML decoding, because the number of time intervals $L$ is now smaller. With the increase in m and $k$, the complexity is increased exponentially. So, the chapter tries to reduce the computational complexity through the proposed chaotic interleaver by eliminating the need of convolutional codes with large constraint lengths [16].

## References

1. Howitt I (2001) WLAN and WPAN coexistence in UL band. IEEE Trans Veh Technol 50(4)
2. Viterbi AJ (1971) Convolutional codes and their performance in communication systems. IEEE Trans Commun Technol 19(5)
3. Bluetooth specifications v1.1. http://www.bluetooth.com
4. Specification of bluetooth system, volume 2, version 1.1, February 22, 2001
5. Bluetooth specifications v1.2. http://www.bluetooth.com
6. Heller JA (1968) Short constraint length convolutional codes. Jet Propul Lab Calif Inst Technol Space Programs Summary 37–54 III:171–177
7. Omura JK (1969) On the Viterbi decoding algorithm. IEEE Trans Inform Theor IT-15
8. Heller JA, Jacobs IM (1971) Viterbi decoding for satellite and space communication. IEEE Trans Commun Technol 19(5)
9. Gilbert E (1960) Capacity of a burst-noise channel. Bell Syst Tech J 39:1253–1266
10. Chen L-J, Sun T, Chen Y-C Improving Bluetooth EDR data throughput using FEC and interleaving
11. Ahiara Wilson C, Chbuisi I (2015) Simulink modeling of convolutional encoders. SSRG Int J Electr Electron Eng (SSRG-IJEEE) 2(5)
12. Arasteh D (2006) Teaching convolutional coding using MATLAB in communication systems course. In: Proceedings of the 2006 ASEE Gulf-Southwest annual conference southern university and A&M College
13. El-Bendary MAM, Abou El-Azm A, El-Fishawy N, Al-Hosarey FSM, Eltokhy MAR, Abd El-Samie FE, Kazemian HB (2012) JPEG image transmission over mobile bluetooth networks with efficient channel coding and interleaving. Int J Electron
14. El-Bendary MAM (2015) Lower complexity of secured WSN networks. Springer (Book Chapter)
15. El-Bendary MAM, Abu El-Azm AF, El-Fishawy NA, Shawky F, El-Samie FE (2011) Embedded Throughput Improving of Low-rate EDR Packets for Lower-latency. In: World Academy of Sciencc, Engineering and Technology, International Journal of Electrical, Computer, Energetic, Electronic and Communication Engineering, 2011
16. Mohsen A, El-Bendary M (2016) Developing security tools of WSN and WBAN networks applications. Springer

# Chapter 6
# Simulation Scenarios of Pseudo-coding Techniques

## 6.1 Introduction

In the previous chapter, the effects of interleaving technique on a performance of Bluetooth system have been discussed. In this chapter, we investigate the performance of the Bluetooth system with using block interleaver and its effects on image transmission. Also, this chapter studies the effects of channel coding on the image with existing interleaving process and without interleaving [1–4].

## 6.2 Interleaving

Interleaving does not add any redundancy to the bit stream, rather it changes the order the bits are transmitted across the channel, such that two consecutive bits are not transmitted consecutively across the channel. The reason for doing this is that whenever a fading occurs or noise burst occurs, it usually covers multiple bit intervals. This leads to a burst of errors, where more than one consecutive bit is in error. Channel coding, whether block coding or convolution coding, is most optimum when the errors occur randomly and not in burst. As shown in previous Chaps. 4 and 5, interleaving can be viewed as a type of time diversity [5–8].

There are two types of interleavers:

1. Block interleavers and
2. Convolution interleavers.

The block interleaver uses a random access memory (RAM) having M rows and N columns as shown in Fig. 6.1.

The input bit stream to the interleaver is written row by row starting with the top row. When all the M x N memory blocks are written to data stored in the RAM, it is


M. A. M. El-Bendary, *Wireless Personal Communications*,
Signals and Communication Technology,
https://doi.org/10.1007/978-981-10-7131-7_6

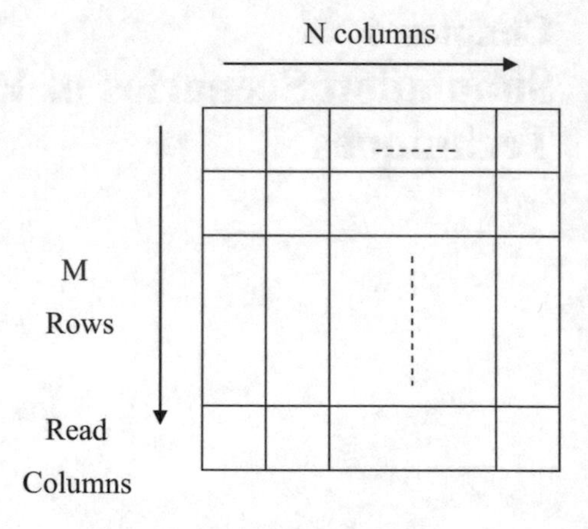


Fig. 6.1 Block interleaver—data reading

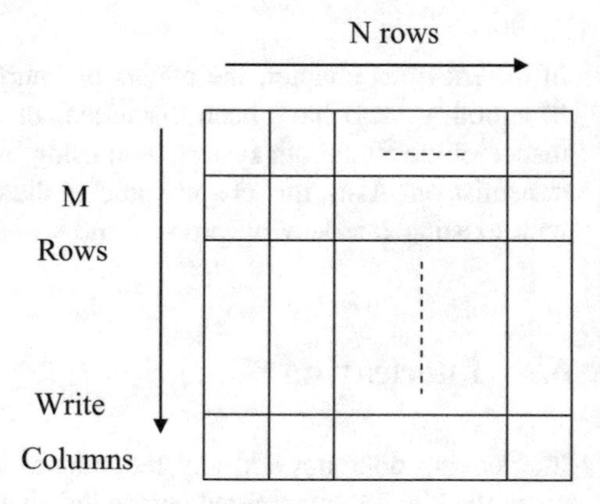


Fig. 6.2 Block de-interleaver—data writing

readout column by column starting with the top element of the left-hand side column [9].

The de-interleaver reverses the interleaving operation. This is done by writing the received data at input of de-interleaver to a RAM having M rows and N columns as shown in Fig. 6.2. The input bits stream to the de-interleaver is written column by column starting with the left-hand side column. When all the M x N memory blocks are written to the data stored in the RAM, it is readout row by row starting with the top row [10–12].

Let us assume that the input bit stream to interleaver is:

$$x_1, x_2, x_3, \ldots x_{12}$$

Let us assume the interleaver has four rows and three columns. The input to the interleaver is written row by row, and hence, they are arranged as follows:

$$\begin{matrix} x_1 & x_2 & x_3 \\ x_4 & x_5 & x_6 \\ x_7 & x_8 & x_9 \\ x_{10} & x_{11} & x_{12} \end{matrix}$$

The outputs from the interleaver are read column by column, hence the output bit stream is:

$$x_1, x_4, x_7, x_{10}, x_2, x_5, x_8, x_{11}, x_3, x_6, x_9, x_{12}$$

A block interleaver operates on discrete blocks with no overlapping between blocks. Another type of interleaving is convolution interleaving. In this type, interleaving is more continuous as there is no clear boundary between one block and the next. In this chapter, we concern block interleaver effects on image with Bluetooth system.

## 6.3 Chaotic Maps

In this section, chaotic maps are discussed and used for image encryption and random data interleaving. A chaotic map is first generalized by introducing parameters and then discretized to a finite square lattice of points, which represent data items. The process of chaotic map can be described as follows [13, 14]:

***Choosing the Basic Map Description***:

In 1998, J. Fridrich shows the difference between the generalized and discretized chaotic map. In this step, the mathematical form of a chaotic two-dimensional map $f$ which maps the unit square $I \times I$ where $I = [0, 1]$, into itself in a one-to-one manner is chosen. There are a number of different chaotic maps, which seem to be suitable for ciphering purposes. However, simple chaotic maps are utilized for ciphering/deciphering performing quickly. The map should allow natural parameterization to create a short ciphering key with a large number of possible keys. Such maps are often described geometrically (e.g., the Baker map, the cat map, the standard map, etc.). In this chapter, the chaotic maps are chosen to act as randomizing/interleaving tool [14].

***Generalization Step***:

In the second step, a set of parameters is introduced into the map to create a part of the ciphering key. If the basic map is described in geometric terms, the parameterization is usually straightforward. If it can be done in several different ways, the one, which best suits, the purpose of secure ciphering needs to be chosen. Two-dimensional chaotic maps will be characterized by a sequence of integers. Another parameter is the number of applications of the chaotic map. It is typically an integer less than 15 [14].

***Discretization Step***:

This step consists of modifying the generalized map to account for the fact that an image is a finite lattice of points. The domain and range of the map are changed from the unit square $I \times I$ to the lattice $N_0^N \times N_0^N$, where $N_0^N = \{0, \ldots N-1\}$ with $N$ equal to the number of pixels in one row. The discretized map $F$ takes each pixel and assigns it to some other pixel in a bijective manner (e.g., the discretized version is a permutation of pixels). The discretization must satisfy the following asymptotic property [14]:

$$\lim_{N\to\infty} \max_{0\le i,j<N} |f(i/N, j/N) - F(i,j)| = 0 \tag{6.1}$$

where $f$ is the continuous basic map and $F$ is the discretized version. The formula requires the discretized map to become increasingly closer to the continuous map as the number of pixels tends to infinity.

### 6.3.1 Two-Dimensional Baker Map

The Baker map, B, is described with the following formulas

$$B(x,y) = (2x, y/2) \quad \text{when} \quad 0<\mathrm{x}<1/2 \tag{6.2}$$

$$B(x,y) = (2x-1, y/2+1/2) \quad \text{when} \quad 1/2<\mathrm{x}<1 \tag{6.3}$$

The map acts on the unit square as depicted in Fig. 6.3. The left vertical column [(0, ½) × (0, 1)] is stretched horizontally and contracted vertically into the rectangle [(0, 1) × (0, ½)] and the right vertical column [(½, 1) × (0, 1)] is similarly mapped into [(0, 1) × (½, 1)]. The Baker map is a chaotic bijection of the unit square $I \times I$ into itself [15].

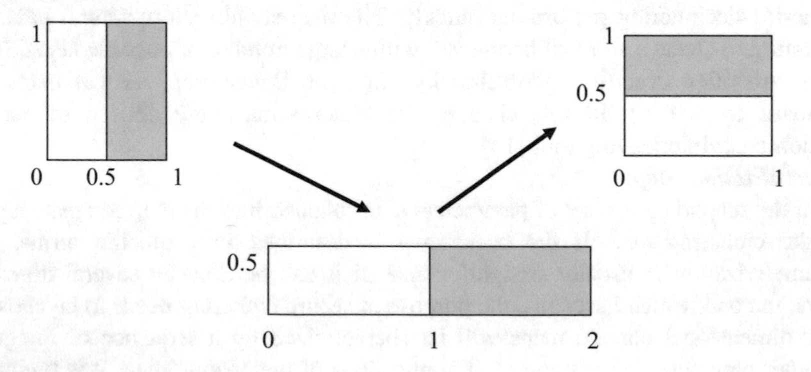


**Fig. 6.3** Baker map—simple example

### 6.3.2 Generalized Baker Map

The map can be generalized in the following way, instead of dividing the square into two rectangles of the same size; the square is divided into $k$ vertical rectangles:

$$[F_{i-1}, F_i]x(0,1) \quad \text{where} \quad i = 1, \ldots, \text{k}.$$

$$F_i = p_1 + \cdots p_i \quad \text{when} \quad F_0 = 0$$

Such as $p_1 + \cdots p_k = 1$, as shown in Fig. 6.4. The lower right corner of the $i$th rectangle is located at $F_i = p_1 + \cdots + p_i$. The generalized Baker map stretches each rectangle horizontally by the factor of $1/p_i$. At the same time, the rectangle is contracted vertically by the factor of $p_i$. Finally, all rectangles are stacked on top of each other as in Fig. 6.4. Formally, it is given by:

$$B(x,y) = \left(\frac{1}{p_i}(x - F_i), p_i y + F_i\right) \text{ for } (x,y) \in [F_i, F_i + p_i) \times [0,1) \tag{6.4}$$

Equation (6.4) is convenient to denote the Baker map and its generalized version as $B(1/2, 1/2)$ and $B_{(p_i,\ldots,p_k)}$, respectively. The generalized map inherits all important properties of the Baker map such as sensitivity to initial conditions and parameters, mixing, and bijectiveness [16].

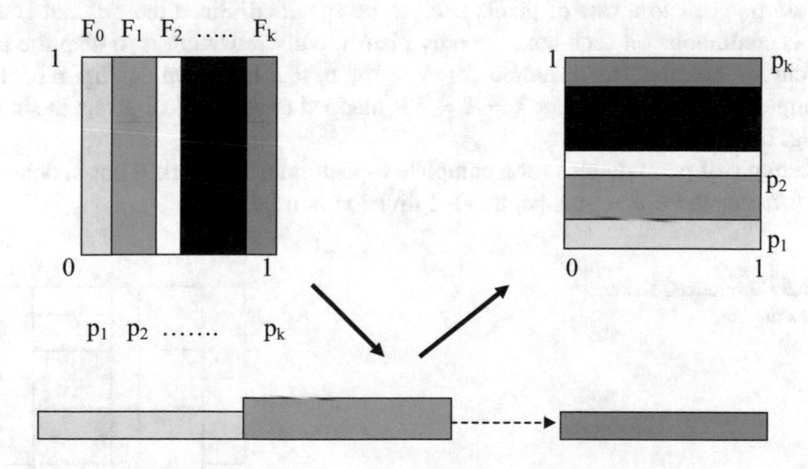


**Fig. 6.4** Generalized Baker map

### 6.3.3 Discretized Baker Map

Version A is a type of discretized map in which any pixel takes its new position according to this map. In particular, the discretized map is required to assign a pixel to another pixel. Since the discretized map is desired to inherit the properties of the continuous basic map, the discretized map should become increasingly close to the basic map as the number of pixels tends to infinity. This requirement is expressed mathematically with Eq. (6.1). The discretized generalized Baker map will be denoted $\mathrm{B}(n_1, \ldots, n_k)$, where the sequence of $k$ integers, $n_{1,\ldots,}n_k$, is chosen such that each integer $ni$ divides $N$, and $n_{1,\ldots,}n_k = N$. Denoting $\mathrm{N_i} = \mathrm{n_1} + \cdots + \mathrm{n_i}$, the pixel $(r, s)$, with $N_i \leq r \leq N_i + n_i$ and $0 \leq s < N$, is mapped to the following equation [17]:

$$B_{(n_1,\ldots,n_k)}(r, s) = \left(\frac{N}{n_i}(r - N_i) + s \bmod \frac{N}{n_i}, \frac{n_i}{N}\left(s - s \bmod \frac{N}{n_i}\right) + N_i\right) \tag{6.5}$$

Equation (6.5) is based on the following geometrical considerations. An $N \times N$ square is divided into vertical rectangles of height $N$ and width $n_i$. Following the action of the generalized Baker map, these vertical rectangles should be stretched in the horizontal direction and contracted in the vertical direction to obtain a horizontal $n_i \times N$ rectangle. To achieve this for the discretized map, each vertical rectangle $N \times n_i$ is divided into $n_i$ boxes $N/n_i \times n_i$ containing exactly $N$ points as shown in Fig. 6.5 [18, 19].

Each of these boxes is mapped to a row of pixels. Since there are $n_i$ boxes, a horizontal rectangle $n_i \times N$ is obtained, as required. Now, how the pixels in each box are mapped to a row of pixels need to be specified. Since the original Baker map is continuous on each box, the only plausible discretization is to map the box column by column. An example for $N = 16$, $n_i = 2$ is shown in Fig. 6.6. The rectangle $N/n_i \times n_i = 16/2 \times 2 = 8 \times 2$ is mapped to a row of 16 pixels as shown in Fig. 6.6.

Example of permutations for a complete 6-pixel image is worked out in detail in Fig. 6.7. For the 6-pixel image, a 3-1-2 division is used [17].

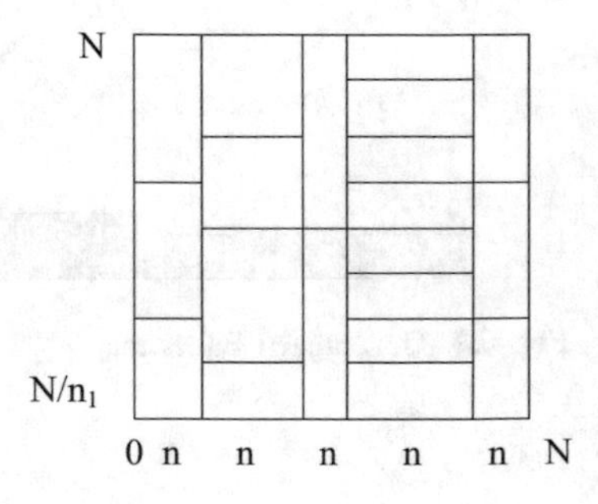


**Fig. 6.5** Discretized Baker map example

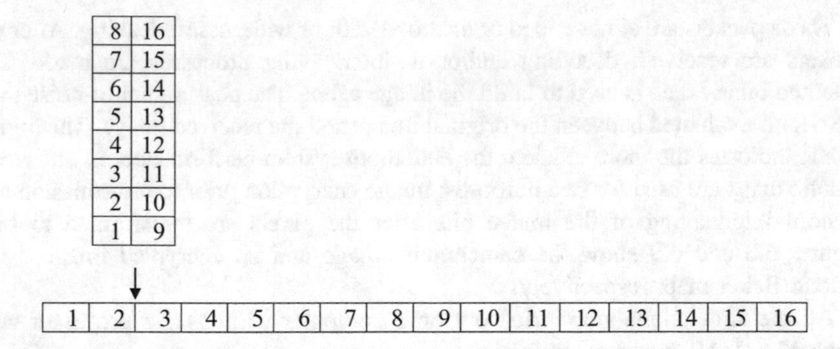


**Fig. 6.6** Non-square Baker map

**Fig. 6.7** Discretized Baker map for a 6-pixel image by (3, 1, 2)

| | | | | | |
|---|---|---|---|---|---|
| 1 | 2 | 3 | 4 | 5 | 6 |
| 7 | 8 | 9 | 10 | 11 | 12 |
| 13 | 14 | 15 | 16 | 17 | 1 |
| 19 | 20 | 21 | 22 | 23 | 24 |
| 25 | 26 | 27 | 28 | 29 | 30 |
| 31 | 32 | 33 | 34 | 35 | 36 |

| | | | | | |
|---|---|---|---|---|---|
| 17 | 11 | 5 | 18 | 12 | 6 |
| 35 | 29 | 23 | 36 | 30 | 24 |
| 34 | 28 | 22 | 16 | 10 | 4 |
| 7 | 1 | 8 | 2 | 9 | 3 |
| 19 | 13 | 20 | 14 | 21 | 15 |
| 31 | 25 | 32 | 26 | 33 | 27 |

The same approach used in image encryption can be used for data randomization if the binary data is represented in a matrix form.

The previous discussion is concluded to be powerful interleaver for data randomizing. The idea of interleaving based on the chaotic Baker map is presented in different researches work such as [20–25]. In this figure, example of the chaotic interleaver Baker map based is cleared as follows.

## 6.4 Simulation Assumptions

All the simulation assumptions used in the previous chapters are maintained in this chapter except that an image is used for transmission rather than a random data. The image which is used in the experiments is 256 × 256 cameraman image. The computer simulation experiments concentrated on uncoded $DH_1$ and encoded $DM_1$ packets only, the chosen packets for the simulation experiments are shorter packets of WPAN-Bluetooth system. The data randomizing techniques perfom better with long packets. In the presented Matlab code, the image of experiments is prepared and processed in steps, it is transformed firstly to binary form and segmented to small packets. Every packets will processed individualy, from the encoding process, interleaving, and modulating to transmitting over the communications channel.

These packets are either coded or uncoded with or without interleaving. After the packets are received, decoding and/or de-interleaving processes are made. The obtained binary data is used to build the image again. The peak signal-to-noise ratio (PSNR) is evaluated between the original image and the received image. The higher PSNR indicates the more efficient the Bluetooth system performance. In our work, chaotic maps are used for two purposes; image encryption prior to transmission and random interleaving of the image bits after the pixels are transformed to bits. Figures 6.8 and 6.9 show the cameraman image and its encrypted image using chaotic Baker map, respectively.

All the simulations are carried out on a computer with 2 GHz processor with using MATLAB program simulation.

**Fig. 6.8** Original cameraman image

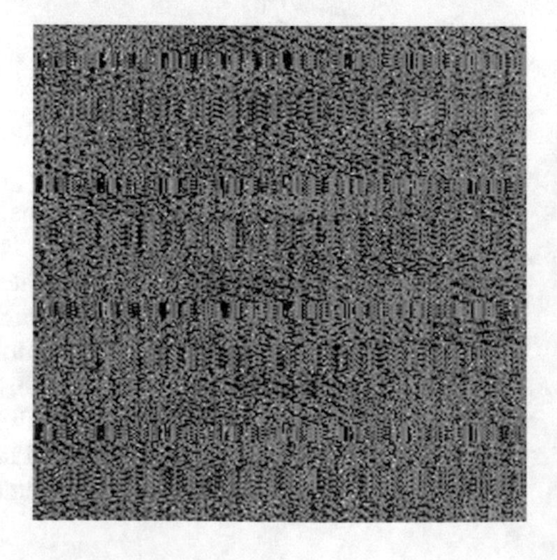

**Fig. 6.9** Encrypted cameraman image

The computer simulation image size is 256 × 256 pixel. The key of image randomizing is related its size, so the proposed key is {(10, 5, 12, 5, 10, 8, 14, 10, 5, 12, 5, 10, 8, 14, 10, 5, 12, 5, 10, 8, 14, 10, 5, 12, 5, 10, 8, 14), all numbers sum=256} with repeated pattern (10, 5, 12, 5, 10, 8, 14). Sum of the repeated pattern numbers is 64, Fig. 6.39 shows the encrypted image version.

In this section, the simulation environment used for comparison between the traditional Bluetooth system and the proposed modifications of this system is described.

The modulation technique used for data transmission in our simulation experiments is the Binary Phase Shift Keying (BPSK) scheme instead of the Gaussian Frequency Shift Keying (GFSK) scheme which is implemented in the standard Bluetooth system. This is attributed to the simplicity of BPSK.

The Monte Carlo simulation method is used in the simulation experiments to compare between the traditional Hamming (15, 10) code used in the standard Bluetooth packet, block interleaving technique, chaotic map encryption, and the proposed coding schemes. This method ensures obtaining correct statistical results.

An important assumption used in the simulation is that a packet is discarded if there is an error in the AC, HD, or PL (after decoding) which was not corrected using the error correction scheme. This is a realistic assumption to simulate the real Bluetooth systems operation.

In this simulation, hard decision is assumed at the receiver in the decoding process for all channel codes. In the simulation, the interference effects are neglected. The packet lengths in all experiments are kept fixed for all coding schemes. This is at the expense of payload lengths.

In some simulation experiments, a block-fading channel is assumed. It is a slow and frequency non-selective channel, where symbols in a block undergo a constant fading effect. The fading coefficients are uncorrelated from block to block regardless of the length of the block. This means that the fading is constant over one hop and independent from hop to hop. Rayleigh fading statistics are also considered. These statistics represent the case of non-line-of-sight links [18].

%##########################

The MATLAB code of the transmitted image after the randomizing process is presented as follows:

Notes:-

- *The randomizing tool is the chaotic Baker map-based interleaving which is studied. The required function codes are presented also.*
- *The randomizing process can be performed packet by packet, and the functions also are presented.*
- *For the encoded scenarios, the reader can back to the previous chapters.*

%############## The Main code ##############

```
clc
clear all
close all
 %_-_-____________________BLUEOOTH SIMULATION OVR DIFFRNT CHANNELS
 % USING THE PESUDO CODING EMPLOYING THE CHAOTIC ENCRYPTION FOR THE
WHOL
 % TARNSMITTED IMAGES OR PACKET-BY-PACKET.

 %THE LENGTH OF BT_PACKET  BEFORE ENCODING PROCESS
 % LENGTH=4+(24+6)+4+18+PAYLOAD
                             %MEDIUM RATE(K)
 %SO THE LENGTH OF  ====> DM1=56+PAY(160) =216
===========>FEC(15,10)HAM.
 %SO THE LENGTH OF  ====> DM3=56+PAY(1000)=1056
===========>FEC(15,10)HAM.
 %SO THE LENGTH OF  ====> DM5=56+PAY(1830)=1886
===========>FEC(15,10)HAM.
                %_______________HIGH RATE(K)______________%

 %SO THE LENGTH OF  ====> DH1=56+PAY(240) =296  ===========> NO FEC
 %SO THE LENGTH OF  ====> DH3=56+PAY(1500)=1556 ===========> NO FEC
 %SO THE LENGTH OF  ====> DH5=56+PAY(2745)=2801 ===========> NO FEC
                %______________AFTER ENCODING_____________%
 % THE LENGTH OF BT_PACKET AFTER ENCODING PROCESS
 %LENGTH= 4+64(SYNC WORD)+4+54(HEADER)+PAYLOAD
                              % MEDIUM RATE(N)
 %SO THE LENGTH OF  ====> DM1=126+PAY(240) =366
<===========FEC(15,10)HAM.
 %SO THE LENGTH OF  ====> DM3=126+PAY(1500)=1626
<===========FEC(15,10)HAM.
 %SO THE LENGTH OF  ====> DM5=126+PAY(2745)=2871
<===========FEC(15,10)HAM.
                 %_______________HIGH RATE(N)__________________%

 %SO THE LENGTH OF  ====> DH1=126+PAY(240) =366  ===========> NO
FEC
 %SO THE LENGTH OF  ====> DH3=126+PAY(1500)=1626 ===========> NO
FEC
 %SO THE LENGTH OF  ====> DH5=126+PAY(2745)=2871 ===========> NO
FEC
 % THE RATE WILL BE AS SHOWN

             % DM1  -------> R1=K1/N1=216/366.
             % DM3  -------> R3=K3/N3=1056/1626.
             % DM5  -------> R5=K5/N5=1886/2871.
             % DH1---------> RH1=K'1/N'1=296/366.
             % DH3  -------> RH3=K'3/N'3=1556/1626.
             % DH5  -------> RH5=K'5/N'5=2801/2871.
```

```
    %HERE WE KNOW RATE OF EACH BLUETOOTH PACKET TRYING TO CHANGE
THE
    %FEC AND COMPARE 1-FIRST USE BLOCK CODES
    %                2-SECOND USE CONV.ECODER WITH PUNCHERING
PROCESS
    %      TO KEEP THE RATE AS MUCH AS WE CAN

    %################### INPUTSDEFINITIONS########################
    %###################                  #######################
    %####################  ACCESS CODE CONV ######################
     %STATE_N_AC =16;%input('STATE NO        STATE_N_DM1  =: ');
     %HALF_ST_AC =8;%input('1/2STATE NO        HALF_ST_DM1  =: ');
     %IN_AC      =1;%input('encoder input        IN_AC  =: ');
     %N_out_AC   =2;%input('encoder output       N_ont_AC =: ');
     %K_AC       =5;%input('the constraint len.        K_AC =: ');
     %L_AC       =30;%input('data length                L_AC =: ');
     %M_AC       =4;%input('TAIL length                L_DM1 =: ');

%################################################################
    %####################  HEADER PART CONV
########################
     %STATE_N_HD =16;%input('STATE NO        STATE_N_DM1  =: ');
     %HALF_ST_HD =8;%input('1/2STATE NO        HALF_ST_DM1  =: ');
     %IN_HD        =1;%input('encoder input        IN_HD  =: ');
     %N_out_HD     =2;%input('encoder output       N_out_HD =: ');
     %K_HD         =5;%input('the constraint len.     K_HD =: ');
     %L_HD         =18;%input('data length              L_HD =: ');
     %M_HD       =4;%input('TAIL length                L_DM1 =: ');

%#################################################################
    segm1=10 ;% input('the no of k data in of encoder =  ');
%segma1=10
    segm2=15;% input('the no of k data in of decoder =  ');
%segma2=15
    n2=63    ;% input('the no of codeword bch-access  =  ');
%n2=64
    k2=30    ;% input('the no of k data bch encoder =  ');     %
k2=30
   % N=input('the NO of iteration                       =  ');
    %G IS GENERATOR MATRIX FOR HAMMING (15,10)ENCODER

    %G=[1 1 0 1 0 1 0 0 0 0 0 0 0 0 0;
       %0 1 1 0 1 0 1 0 0 0 0 0 0 0 0;
       %1 1 1 0 0 0 0 1 0 0 0 0 0 0 0;
       %0 1 1 1 0 0 0 0 1 0 0 0 0 0 0;
       %0 0 1 1 1 0 0 0 0 1 0 0 0 0 0;
       %1 1 0 0 1 0 0 0 0 0 1 0 0 0 0;
       %1 0 1 1 0 0 0 0 0 0 0 1 0 0 0;
       %0 1 0 1 1 0 0 0 0 0 0 0 1 0 0;
       %1 1 1 1 1 0 0 0 0 0 0 0 0 1 0;
       %1 0 1 0 1 0 0 0 0 0 0 0 0 0 1];

      % H  IS PARITY_CHECK MATRIX FOR HAMMING (15,10) DECODER
```

```
%H=[1 0 0 0 0 1 0 1 0 0 1 1 0 1 1;
   % 0 1 0 0 0 1 1 1 1 0 1 0 1 1 0;
    %0 0 1 0 0 0 1 1 1 1 0 1 0 1 1;
    %0 0 0 1 0 1 0 0 1 1 0 1 1 1 0;
    %0 0 0 0 1 0 1 0 0 1 1 0 1 1 1];


snr=[0:5:30];
y=length(snr);
snrabs=[];
counter_iter=0;
Y=0;
for j=1:y

    snrabs(j)=10^(snr(j)/10);
    sigma=1/sqrt(2*snrabs(j));
    error_dh1=0;
    error_dm1=0;
    error_dh3=0;
    error_dm3=0;
    error_dh5=0;
    error_dm5=0;
    per1=0;
    per2=0;
    per3=0;
    counter_iter=counter_iter+1

    %DEEP FADING


p2=round(rand(1,30));               %sync word LAP=24+6

p4=round(rand(1,18));               %header
%#################################################################
%#################################################################
%######################     IMAGE TO BINARY - MANCAMERA
##############
  f=imread('cameraman.tif');
    f=double(f);
    x=randomization(f);

 [M,N]=size(x);
  g=im2col(x, [M,N], [M,N], 'distinct');
```

```
%3###############################################################
#
     %3################################


     R=4
      h=dec2bin(double(g));
      UU=5

   %##############################################################
   %################   IMAGE TO  VECTOR   ########################

    for i=1:8
         frame(1,(i-1)*(256^2)+1:i*256^2)=h(:,i);
    end
    %####################################
     %#############################################################
    %################  DIVIDING THE FRAME BT PL  ##################


    x=0;
for ff=1:2048  % CAMERAMAN
%for ff=1:8192    %LENA
    x=x+1
    PL(1,1:256)=frame(1,(ff-1)*256+1:ff*256);

  %#########################################################
  %#########################################################
  %################  CONVERSION  ####################
          %--------------------------------=
              for w=1:256
                PL_BT_form(w)=str2num(PL(w));

              end

     %########################################################
              delay=[0 1 2];     %set DELAYS OF THREE SHIFT
REGISTER

    for r=1:6
        PL_BT_form(256+r)=0;
    end
    [PL_BT_intcon,statey]=muxintrlv(PL_BT_form,delay);
    %###############  BLUETOOTH SYSTEM   #####################
    %#########################      ########################

               %--------------------------------=

    bt_pkth1=[p2,p4,PL_BT_intcon];
    %bt_pkth3=[p2,p4,dh3];
    %bt_pkth5=[p2,p4,dh5];
```

```
%****************************************************************

    % p2 sync word use code bch code(64,30)

    %p2=gf(p2);              %BCH CODE ACCESS CODE PART IN GENERAL
FORM
    %[genpoly,t]=bchgenpoly(n2,k2);

     %pg = bchpoly(n2, k2);
     p2_encode=bchenco(p2,n2,k2);

      %#########  CONV  ENCODER

%[enc_word_AC,sym_AC,data_len_AC]=conv_encode_AC(STATE_N_AC,M_AC,
     %IN_AC,N_out_AC,K_AC,L_AC,y_AC);
    %*************************************
    % P4 IS HEADER USE REPETITION CODE 1/3
    for k=1:18
        for l=1:3
            p4_encode(3*(k-1)+l)=p4(k);
        end
    end


    %-----------------------------------
    bt_pkt_ench1=[p2_encode,p4_encode,PL_BT_intcon];


    %------------------------------


    mod_bt_pkth1=2*bt_pkt_ench1-1;
    %mod_bt_pkth3=2*bt_pkt_ench3-1;

    %mod_bt_pkth5=2*bt_pkt_ench5-1;

    %**********************CHANNEL EFFECTS

    %rec_bt_pkth1=mod_bt_pkth1+sigma*randn(1,length(bt_pkt_ench1));
    %rec_bt_pkth3=mod_bt_pkth3+sigma*randn(1,length(bt_pkt_ench3));
    %rec_bt_pkth5=mod_bt_pkth5+sigma*randn(1,length(bt_pkt_ench5));

    %---------------------NO FEC IN PAYLOAD-----------------

    %here we will add rayleigh fading with perfect interleaver

    deep_fad=sqrt((0.5*((randn(1,1))^2))+(0.5*((randn(1,1))^2)));
    bt_pkth1_d_fad1=mod_bt_pkth1*deep_fad;

rec_bt_pkth1=bt_pkth1_d_fad1+sigma*randn(1,length(bt_pkt_ench1));
    %2@@@@@@@@@

    %bt_pkth3_d_fad3=mod_bt_pkth3*deep_fad;
```

```
%rec_bt_pkth3=bt_pkth3_d_fad3+sigma*randn(1,length(bt_pkt_ench3));
    %2@@@@@@@@@

    % bt_pkth5_d_fad5=mod_bt_pkth5*deep_fad;

%rec_bt_pkth5=bt_pkth5_d_fad5+sigma*randn(1,length(bt_pkt_ench5));
     %--------------------  CORRELATED FADING CHANNEL  -----------
------%
    %[row,column]=size(bt_pkt_ench1);
        %a=jack_fading(column);

%rec_bt_pkth1=a.*(mod_bt_pkth1)+sigma*randn(1,length(bt_pkt_ench1))
;
    %**********************DEMODULATION

    %demod_bt_pkt1=rec_bt_pkt1>0;
    %demod_bt_pkt3=rec_bt_pkt3>0;
    %demod_bt_pkt5=rec_bt_pkt5>0;

    %---------------------
    for r=1:length(rec_bt_pkth1)
        if rec_bt_pkth1(r)>0

    demod_bt_pkth1(r)=1;
        else
    demod_bt_pkth1(r)=0;
        end
        end
    %demod_bt_pkth3=rec_bt_pkth3>0;
    %-__________________________________


    %***********************DECODING EACH PART ALONE



    %NOW HAMM CODE FOR DM PKT BUT DIRECT FOR DH PKT

    %p5_dec1=decoding(p5_rec1,H,segm2);%IT IS FUNCTION  H IS
PARITY- CHECK
    %MATRIX

    % so the pkt after THAT IS
    %bt_pkt_dec1=[p1_dec1,p2_dec1,p3_dec1,p4_dec1,p5_dec1];
%<=============
```

```
%************dh dh dh dh dh1
%############################## DH1
 p2_rech1=demod_bt_pkth1(1,1:63);   % sync word DM1
 p4_rech1=demod_bt_pkth1(1,64:117);
 PL_rech1=demod_bt_pkth1(1,118:length(bt_pkt_ench1));
%############################## DH3


%########################################

%########################################

%p1_dech1=p1_rech1;
 p2_dech1=bchdeco(p2_rech1,k2,6);
 %p3_dech1=p3_rech1;

 % ################

  %p4 IS HEADER REPETION CODE
 %
 for k=1:18
     counter=0;
     for l=1:3
         if p4_rech1(3*(k-1)+l)==1
             counter=counter+1;
         end
     end
     if counter >1
         p4_dech1(k)=1;
     else
         p4_dech1(k)=0;
     end
 end



  %##########################

       PL_dech1=PL_rech1;    %DH1  ONE SLOT
   %#########################################



      %***@#@SO THE FINAL STEP IS ERROE CALCULATION@#@#****

      bt_pkt_dech1=[p2_dech1,p4_dech1,PL_dech1];%DH1
      %bt_pkt_dech3=[p2_dech3,p4_dech3,p5_dech3];%DH3
      %bt_pkt_dech5=[p2_dech5,p4_dech5,p5_dech5];%DH5

%##############################################################
```

```
%##############################################################

        x1=0;
        x2=0;
        x3=0;
        x1=sum(xor(bt_pkt_dech1,bt_pkth1));
        %x2=sum(xor(bt_pkt_dech3,bt_pkth3));
        %x3=sum(xor(bt_pkt_dech5,bt_pkth5));
          error_dh1=error_dh1+x1;
          %error_dh3=error_dh3+x2;
          %error_dh5=error_dh5+x3;
          if x1>0
              per1=per1+1;
          end

          %if x2>0
              %per2=per2+1;
          %end
          %if x3>0
              %per3=per3+1;
          %end


         ber_dh1(j)=error_dh1/(length(bt_pkth1)*(2048*304));
         %ber_dh3(j)=error_dh3/((length(bt_pkth3))*N);
         %ber_dh5(j)=error_dh5/((length(bt_pkth5))*N);
         per_dh1(j)=per1/2048;
         %per_dh3(j)=per2/N;
         %per_dh5(j)=per3/N;


%##########################################################
           %######################## CONV INTER ##################
         PL_deintcon1=muxdeintrlv(PL_dech1,delay);
         PL_deintcon=PL_deintcon1(1,7:length(PL_deintcon1));
          %###################
                  for i=1:256
                        PL_PIC_form(i)=num2str(PL_deintcon(i));

                 end
         %##############################################################

         frame_rec(1,(x-1)*256+1:x*256)=PL_PIC_form;

end

        %##############################################################
        %##############################################################
         %###############3   BACK TO IMAGE FORM    ################

     for yyy=1:8

          pic_rec(:,yyy)=frame_rec(1,(yyy-1)*(256^2)+1:yyy*256^2);

         end
```

```
%##########################################################
%##########################################################

%##############          BACK       ########################
%##############          BACK       ########################

bendo=bin2dec(pic_rec);
bendo_back=col2im(bendo, [M,N], [M,N], 'distinct');
%################### REGENERATION OF IMAGE
LENA###################3

Y=Y+1;

f_rec=derandomization(bendo_back);

imwrite(f_rec/255,'DH1_CHAOTIC
RANDAMIZATION_CONV_INT_FADI','tif');

figure(Y+20)
imshow('DH1_CHAOTIC RANDAMIZATION_CONV_INT_FADING');
output_image=f_rec/255;
   %#######BLOCK INERLEAV WITHOUT CHAOTIC########
   MSE1=sum(sum((double(f)/255-
output_image).^2))/prod(size(f))
    PSNR(j)=(10*log(1/MSE1))/log(10)

end

figure(1)
imshow(x/255);
figure(2)
imshow(bendo_back/255);
figure(4);
semilogy(snr,PSNR,'s-k');
title('SNR TO PSNR FADIDH1_CHAOTIC_RANDOMIZATION_CONV_INT');
xlabel('Eb/No (dB)');
ylabel('PSNR');
grid on;
```

The required Matlab Functions:

```
# The x=randomization(f) Matlab Code #######
 For the whole processed image
function x=randomization(f)
[M,N]=size(f) ;
g=im2col(f, [M,N], [M,N], 'distinct');
h=dec2bin(double(g));
[M1,N1]=size(h);
z=zeros (M1,N1) ;
for i=1:M1
for jj=1:N1
z(i,jj)=str2num(h(i,jj));
end;
end;
      h_chaot = zeros(size(z));

   for j=1:8
       HH=z(:,j);
       FF=reshape(HH,M,N);
       n =
[10,5,12,5,10,8,14,10,5,12,5,10,8,14,10,5,12,5,10,8,14,10,5,12,5
      ,10,8,14];
       %n=[4,1,5,6];
       [pr,pc] = chaomat(n);
       pim = chaoperm(FF,pr,pc,3,'forward');
       h_chaot(:,j)=reshape(pim,M*N,1);
   end


   for i=1:M1
for jj=1:N1
zn(i,jj)=num2str(h_chaot(i,jj));
end;
end;

hn=bin2dec(zn);
x=col2im(hn, [M,N], [M,N], 'distinct');
```

# As shown the length of secret key equals the processed data dimensions.

In the below, the function f=derandomization(x)

Matlab code of the de-randomization Function:-

```
#%############################################
```

```
function f=derandomization(x)
[M,N]=size(x) ;
g=im2col(x, [M,N], [M,N], 'distinct');
h=dec2bin(double(g));
[M1,N1]=size(h);
z=zeros (M1,N1) ;
for i=1:M1
for jj=1:N1
z(i,jj)=str2num(h(i,jj));
end;
end;
      h_rec = zeros(size(z));

   for j=1:8
       HH=z(:,j);
       FF=reshape(HH,M,N);
       n =
[10,5,12,5,10,8,14,10,5,12,5,10,8,14,10,5,12,5,10,8,14,10,5,12,5,10
,8,14];
       %n=[4,1,5,6];
       [pr,pc] = chaomat(n);
       pim = chaoperm(FF,pr,pc,3,'backward');
       h_rec(:,j)=reshape(pim,M*N,1);
   end

   for i=1:M1
for jj=1:N1
zn(i,jj)=num2str(h_rec(i,jj));
end;
end;

hn=bin2dec(zn);
f=col2im(hn, [M,N], [M,N], 'distinct');
%###############################################
```

– In the previous functions, there are two important functions, which are the engine of the randomizing process.

```
#%#########Function#1
```

```
## function [pr,pc]=chaomat(n)
function [pr,pc]=chaomat(n)
%
I=sum(n);
k=size(n,2);
for i=1:k
    N(i+1)=1;
    for j=1:i
        N(i+1)=N(i+1)+n(j);
    end
end
N(1)=1;
%N(0)=1;
for cb=1:k
   for rb=1:n(cb)
      rbstartcol(rb)=mod((rb-1)*I,n(cb));
      rbendcol(rb)=mod((rb*I-1),n(cb));
      rbstartrow(rb)=fix(((rb-1)*I)/n(cb));
      rbendrow(rb)=fix((rb*I-1)/n(cb));
      mincol(rb)=min([rbendcol(rb)+1,rbstartcol(rb)]);
      maxcol(rb)=max([rbendcol(rb),rbstartcol(rb)-1]);
   end

   for i=1:I
      for j=N(cb):N(cb+1)-1
         newindex(i,j-N(cb)+1)=(i-1)*n(cb)+(n(cb)-j+N(cb)-1);
         newindexmod(i,j-N(cb)+1)=mod(newindex(i,j-N(cb)+1),n(cb));
         newindexquotient(i,j-N(cb)+1)=fix(newindex(i,j-
N(cb)+1)/n(cb));
         rowblockindex(i,j-N(cb)+1)=fix(newindex(i,j-N(cb)+1)/I)+1;
      end
   end

   for i=1:I
      for j=1:n(cb)
         for rb=1:n(cb)
            if rowblockindex(i,j)==rb;
               if newindexmod(i,j)>maxcol(rb)
                  col=rbendrow(rb)-newindexquotient(i,j)+(n(cb)-1
                  -newindexmod(i,j))*(rbendrow(rb)-rbstartrow(rb));
               elseif newindexmod(i,j)>=mincol(rb)
                  & newindexmod(i,j)<=maxcol(rb)
                    if rbstartcol(rb)>rbendcol(rb)
                       c=0;
                       d=-1;
                    else
                       c=1;
                       d=1;
                    end
```

```
                        col=(rbendrow(rb)-rbstartrow(rb))*(n(cb)-1-
maxcol(rb))
                        +(rbendrow(rb)-
newindexquotient(i,j)+c)+(maxcol(rb)
                        -newindexmod(i,j))*(rbendrow(rb)-
rbstartrow(rb)+d);
                   else %if newindexmod(i,j)<=mincol(rb)
                      col=I-mincol(rb)*(rbendrow(rb)-rbstartrow(rb))
                      +(rbendrow(rb)-
newindexquotient(i,j)+1)+(mincol(rb)
                      -1-newindexmod(i,j))*(rbendrow(rb)-
rbstartrow(rb));
                   end

                   row=1+I-N(cb+1)+rowblockindex(i,j);
                 end
              end

              pr(i,j+N(cb)-1)=row;
              pc(i,j+N(cb)-1)=col;
          end
      end
  end
```

The second function:-

```
function out=chaoperm(im,pr,pc,num,forward)
%
[rows,cols] = size(im);
mat = zeros([rows,cols,num+1]);
mat(:,:,1) = im(:,:);

for loc=2:num+1
    if(strcmp(forward,'forward'))
        for i=1:rows
            for j=1:cols
                mat(pr(i,j),pc(i,j),loc) = mat(i,j,loc-1);
            end
        end
    elseif(strcmp(forward,'backward'))
        for i=1:rows
            for j=1:cols
                mat(i,j,loc) = mat(pr(i,j),pc(i,j),loc-1);
            end
        end
    end
end
out = mat(:,:,num+1);
```

- So, these presented MATLAB codes of the uncoded packets transmission.
  - For the packet-by-packet scenario, the length of packets will determine the length of secret key. In the following, the function of randomization and de-randomization MATLAB code for another simulation scenarios.
- The secret key is the engine of the randomizing process. Also, the effectiveness of this process depends on the good choice of the key.
- There is the random key, which contains any numbers, and there is the systematic key, where all the key numbers are equal. It is called in this case as the general form of the key.
- As shown in the following:

# Randomizing function:-

```
function [y,m,n]=randomize(x)
[m,n]=size(x);
y=reshape(x,sqrt(m*n),sqrt(m*n));
n1
=[sqrt(n*m)/8,sqrt(n*m)/8,sqrt(n*m)/8,sqrt(n*m)/8,sqrt(n*m)/8,sqrt(
n*m)/8,sqrt(n*m)/8,sqrt(n*m)/8];
%    Case of 256/8   X   8 Time
%   Case of    K=[Sum256] random numbere] Iteration 3 Short
%n=[256,256]
[pr,pc] = chaomat(n1);
y = chaoperm(y,pr,pc,3,'forward');
y=reshape(y,m,n);
```

# De-Randomizing function:-

```
function x=derandomize(y)
[m,n]=size(y);
x=reshape(y,sqrt(m*n),sqrt(m*n));
n1
=[sqrt(n*m)/8,sqrt(n*m)/8,sqrt(n*m)/8,sqrt(n*m)/8,sqrt(n*m)/8,sqrt(
n*m)/8,sqrt(n*m)/8,sqrt(n*m)/8];

[pr,pc] = chaomat(n1);

x = chaoperm(x,pr,pc,3,'backward');
x=reshape(x,m,n);
```

## 6.5 Simulation Results

In this section, several experiments are carried out for the purpose of comparison between the proposed coding schemes, non-interleaving, block interleaving, chaotic map encryption, and the traditional coding scheme in Bluetooth systems. The simulation parameters described in the previous section are used.

The first and second experiments simulate the transmission of cameraman image through ($DH_1$) and uncoded ($DH_1$) packets with using block interleaving technique over an AWGN channel, respectively. The results of these experiments shown in Figs. 6.11 and 6.13 indicate that the $DH_1$ packets with block interleaving are preferred to the $DH_1$ packets at high SNR. In case of non-interleaving, Figs. 6.10 and 6.12 show the cameraman image at SNR = 0 and 5 dB, respectively. But of block interleaving, Figs. 6.11 and 6.13 show the cameraman image at SNR = 0 and 5 dB, respectively. Also, Figs. 6.14 and 6.15 give the PSNR of the received image with the channel SNR variations.

The first two experiments are repeated for a Rayleigh-flat fading channel rather than an AWGN channel. The results of PSNR to SNR are shown in Figs. 6.20 and 6.21. In case of non-interleaving, Figs. 6.16 and 6.18 show the cameraman image at SNR = 5 and 25 dB, respectively. But of block interleaving, Figs. 6.17 and 6.19 show the cameraman image at SNR = 5 and 25 dB, respectively. Effect f block interleaving on $DH_1$ performance appears at High SNR and over fading channel more than AWGN channel.

In the following experiments, the performance of the encoded packets is evaluated with utilizing the presented transmission scenario. These experiments are carried out to test the performance of the proposed chaotic map encryption which is used to encrypt $DH_1$ (uncoded packet) over an AWGN channel and comparing it to the traditional Hamming (15, 10) coding scheme for $DM_1$. The results of PSNR to SNR are shown in Figs. 6.26 and 6.27, respectively. In case of encrypted image

**Fig. 6.10** Image with no interleaving ($DH_1$), AWGN (SNR = 0)

**Fig. 6.11** Image with block interleaving ($DH_1$), AWGN (SNR = 0)

**Fig. 6.12** Image with no interleaving ($DH_1$) AWGN (SNR = 5)

**Fig. 6.13** Image with block interleaving ($DH_1$) AWGN (SNR = 5)

(uncoded packets), Figs. 6.22 and 6.24 show the cameraman image at SNR = 0 and 5 dB, respectively. But of standard $DM_1$, Figs. 6.23 and 6.25 show the cameraman image at SNR = 0 and 5 dB, respectively. Effect of chaotic map encryption on $DH_1$ performance over AWGN channel is ineffective compared to encoded packets.

The previous experiments are repeated over Rayleigh-flat fading channel. The results of PSNR to SNR are shown in Figs. 6.32 and 6.33. In case of encrypted image (uncoded packets), Figs. 6.28 and 6.30 show the cameraman image at SNR = 5 and 25 dB, respectively. But of standard $DM_1$, Figs. 6.29 and 6.31 show the cameraman image at SNR = 5 and 25 dB, respectively. As shown in figures, effect of chaotic map encryption over fading channel is better than the encoded packets at High SNR and also there are no redundant bits.

Another FEC scheme is employed in the following experiments. These groups of the computer simulation experiments are carried out to test the performance of the proposed coding scheme which is BCH (15, 7) code over an AWGN and Rayleigh-flat fading channel, respectively. The results PSNR to SNR are shown in Figs. 6.38 and 6.39. Figures 6.34 and 6.36 show the cameraman image with the proposed encoded packet transmission at SNR = 0 and 5 dB over AWGN channel,

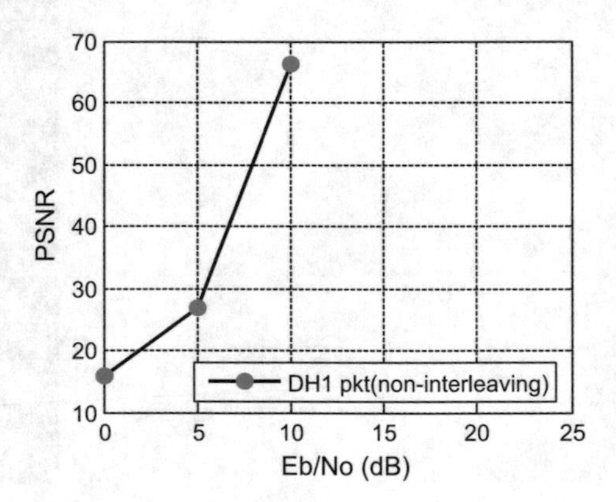


**Fig. 6.14** Image with no interleaving ($DH_1$) AWGN (PSNR to SNR)

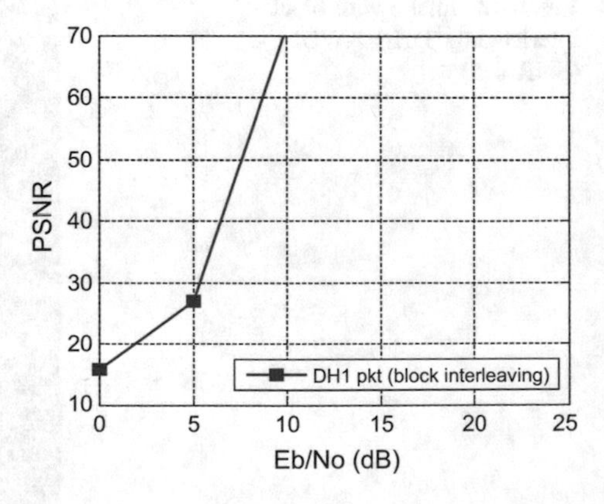


**Fig. 6.15** Image with block interleaving ($DH_1$) AWGN (PSNR to SNR)

**Fig. 6.16** Image with no interleaving ($DH_1$) fading (SNR = 5)

**Fig. 6.17** Image with block interleaving ($DH_1$) fading (SNR = 5)

**Fig. 6.18** Image with no interleaving ($DH_1$) fading (SNR = 25)

**Fig. 6.19** Image with block interleaving ($DH_1$) fading (SNR = 25)

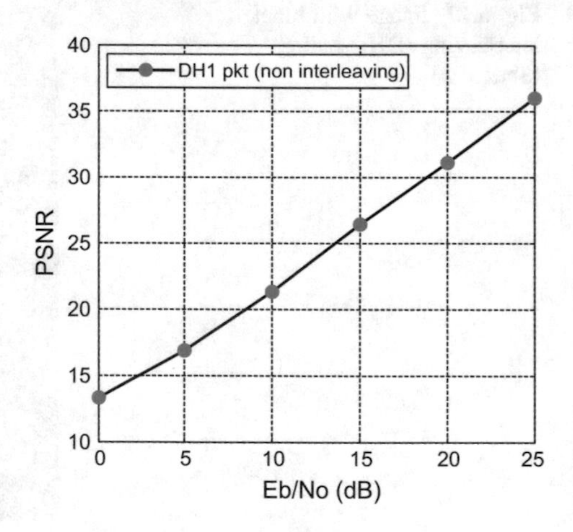


**Fig. 6.20** Image with no interleaving ($DH_1$) fading channel (PSNR to SNR)

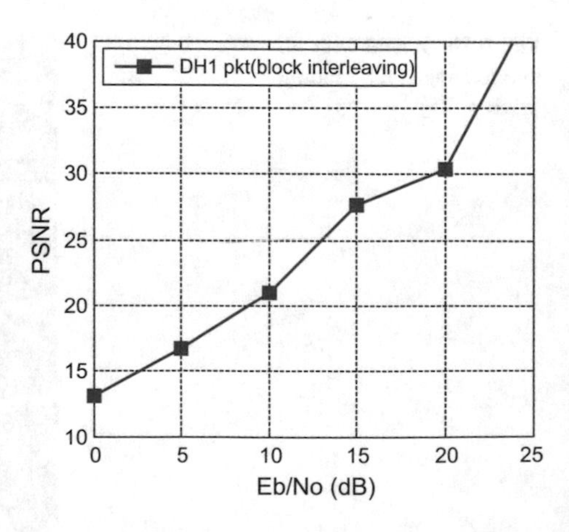


**Fig. 6.21** Image with block interleaving ($DH_1$) fading channel (PSNR to SNR)

**Fig. 6.22** Image with chaotic interleaving over AWGN channel (SNR = 0)

**Fig. 6.23** Image with $DM_1$_H (15, 10) over AWGN channel (SNR = 0)

**Fig. 6.24** Image with chaotic interleaving over AWGN (SNR = 5)

**Fig. 6.25** Image with $DM_1$_H (15, 10) over AWGN (SNR = 5)

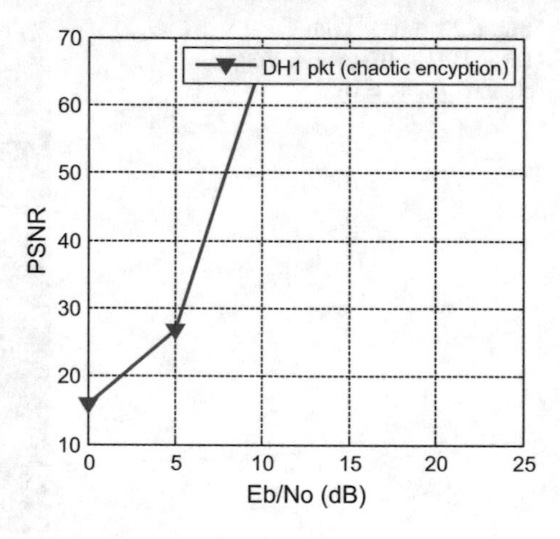


**Fig. 6.26** Image with chaotic interleaving over AWGN channel (PSNR to SNR)

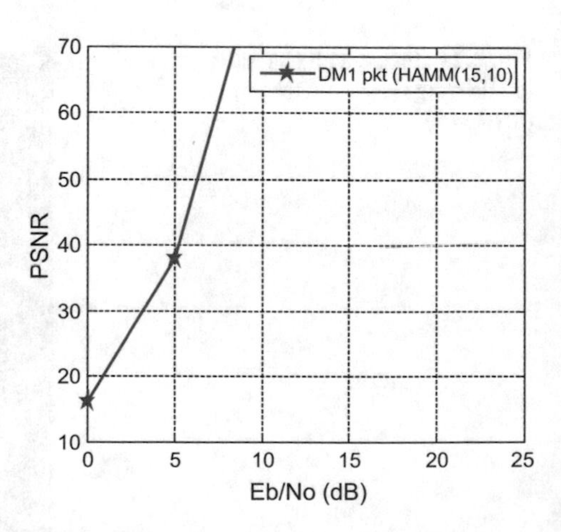


**Fig. 6.27** Image with $DM_1$_H (15, 10) over AWGN channel (PSNR to SNR)

**Fig. 6.28** Image with chaotic interleaving over fading (SNR = 5)

**Fig. 6.29** Image with DM_H (15, 10) over fading (SNR = 5)

**Fig. 6.30** Image with chaotic int. over fading (SNR = 25)

**Fig. 6.31** Image with DM_H (15, 10) over fading (SNR = 25)

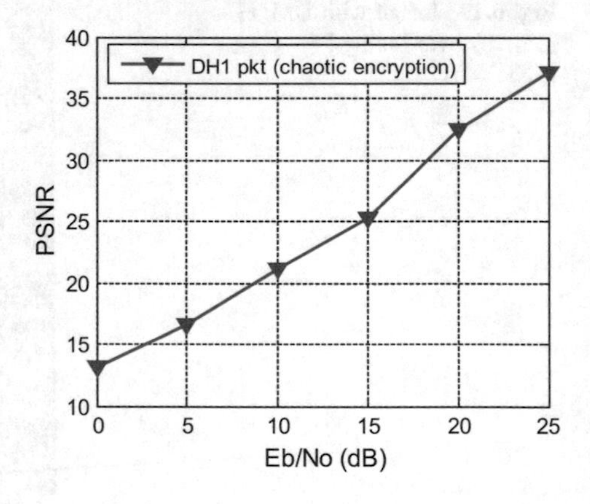


**Fig. 6.32** Image with chaotic int. over fading channel (PSNR to SNR)

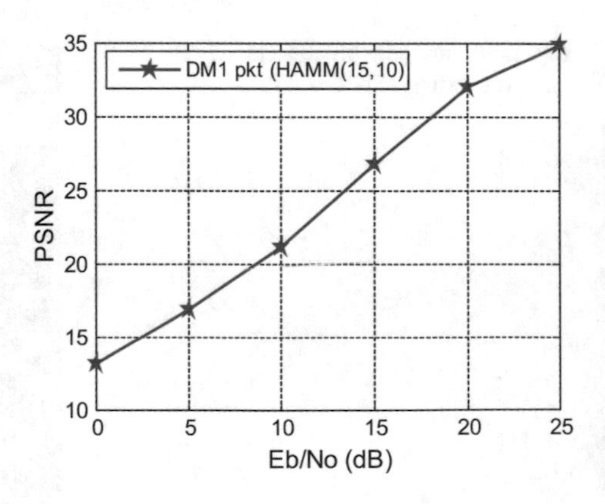


**Fig. 6.33** Image with $DM_1$_H (15, 10) over fading channel (PSNR to SNR)

**Fig. 6.34** Image with DM_BCH (15, 7) over AWGN (SNR = 0)

**Fig. 6.35** Image with DM_BCH (15, 7) over fading channel (SNR = 5)

**Fig. 6.36** Image with DM_BCH (15, 7) over AWGN channel (SNR = 5)

**Fig. 6.37** Image with DM_BCH (15, 7) over fading channel (SNR = 25)

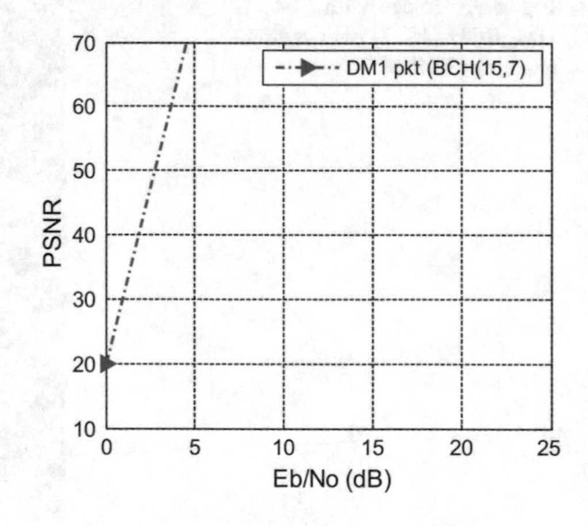


**Fig. 6.38** Image with DM_BCH (15, 7) over AWGN channel (PSNR to SNR)

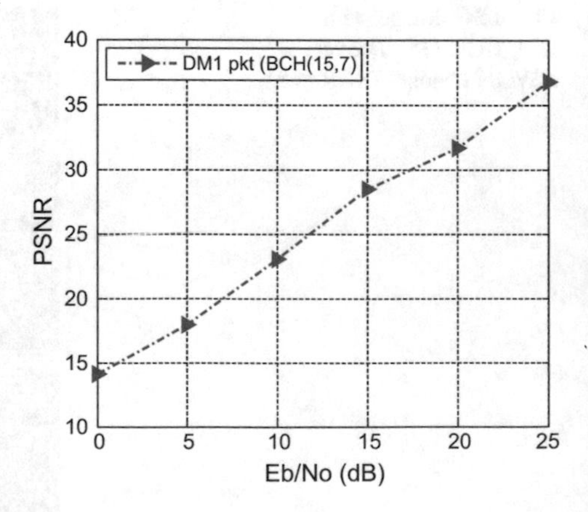


**Fig. 6.39** Image with DM_BCH (15, 7) over fading channel (PSNR to SNR)

respectively. In the proposed encoded packet $DM_1$, Figs. 6.35 and 6.37 show the cameraman image at SNR = 5 and 25 dB over fading channel, respectively.

A comparison study between the standard Hamming (15, 10) code, uncoded packet, and the proposed cases is given in the following figures.

A comparison between all cases over AWGN channel is given in Fig. 6.40. As shown in this figure, over AWGN channel encoded packets perform better than other cases.

Figure 6.41 gives a comparison between different cases of uncoded packets over Rayleigh-flat fading channel. At low SNR, non-interleaving, block interleaving, and chaotic map encryption give the same performance. But at high SNR chaotic map, encryption performs better than standard uncoded packets.

Figure 6.42 shows the error control scheme effects on the image quality by the comparison between the Hamming and BCH codes.

Figure 6.43 shows the performance variation between the different image transmission scenarios with the employing different error control coding or not and using the encryption tools or not.

These comparisons show that the best performance is obtained in the case of $DM_1$ packets implementing the BCH (15, 7) code if the data is transmitted over an AWGN channel.

If the data is transmitted over an interleaving channel, the scenario of $DM_1$ packets implementing the BCH (15, 7) code at low SNR values performs better than other encoding scenarios and to use $DH_1$ packets with block interleaving at high SNR values.

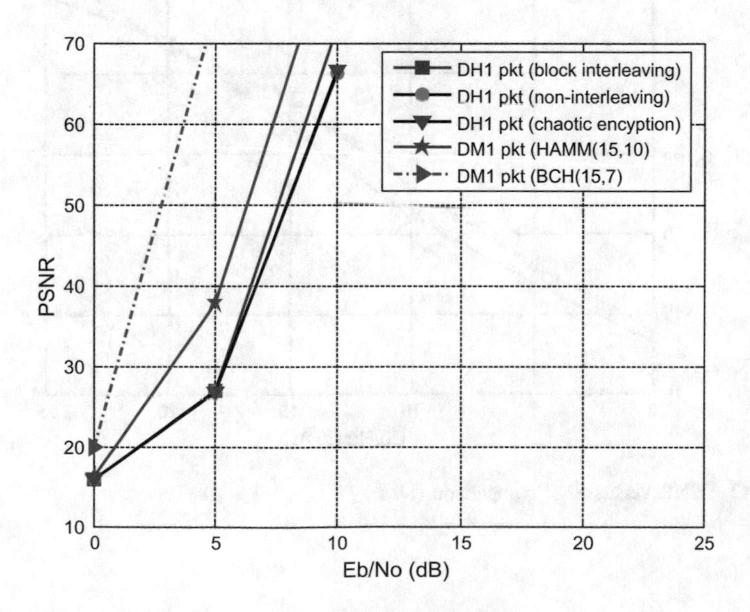


**Fig. 6.40** PSNR versus SNR comparison over AWGN channel

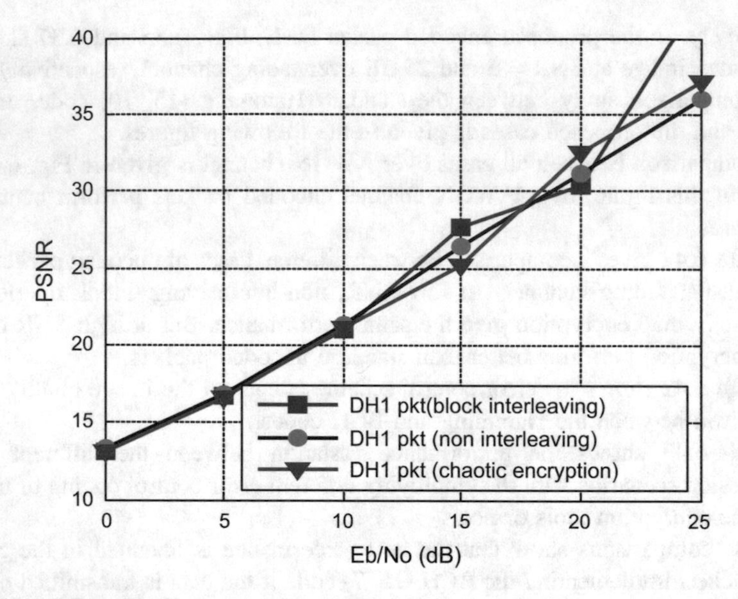


**Fig. 6.41** PSNR versus SNR comparison (fading channel)

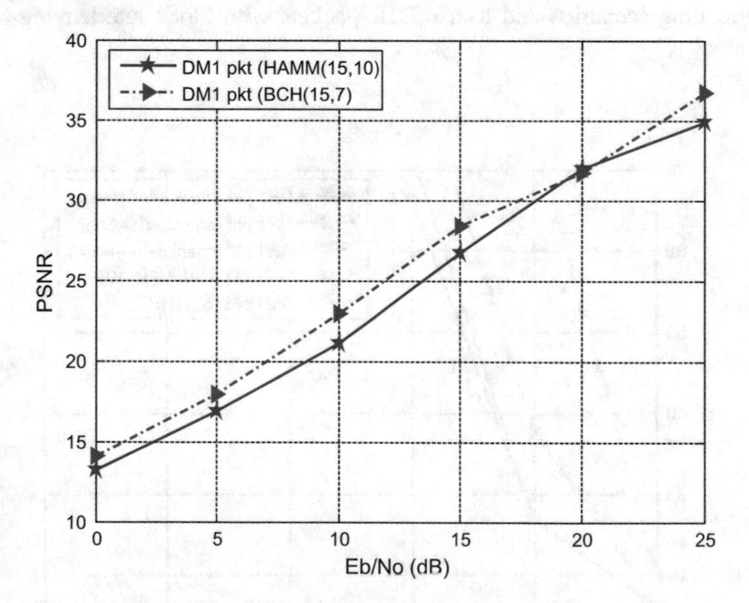


**Fig. 6.42** PSNR versus SNR comparison ($DM_1$)

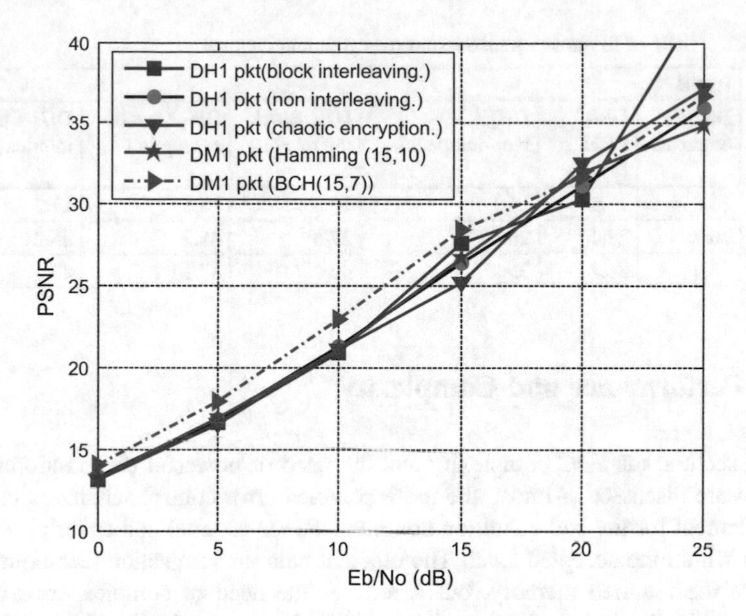


**Fig. 6.43** PSNR versus SNR comparison over fading channel ($DM_1$ and $DH_1$ packets)

Table 6.1 gives a comparison between the standard and proposed cases over an AWGN channel.

Tables 6.1 and 6.2 give a tabulation of the results of the previous experiments. Table 6.1 tabulates the simulation results in the case of an AWGN channel. Also, Table 6.2 gives a tabulation of the simulation results in the case of a fading channel.

These simulation experiments reveal the following remarks:

Over an AWGN channel, encoded packets perform better than uncoded packets in all cases. At low SNR values, $DM_1$ packets perform better than uncoded $DH_1$ packets using chaotic interleaving. On the other hands at high SNR values, chaotic interleaving is better over Rayleigh-flat fading channel.

Chaotic interleaving in uncoded $DH_1$ packets performs better than block interleaving. Over Rayleigh-flat fading channels, $DH_1$ packets implementing the block interleaving technique perform better than the chaotic encryption technique.

**Table 6.1** PSNR of the image transmission over an AWGN channel

| SNR (dB) | PSNR | | | | | |
|---|---|---|---|---|---|---|
| | $DM_1$ Standard | $DM_1$ BCH(15, 7) | $DH_1$ Non-interleaving | $DH_1$ Block interleaving | $DH_1$ Chaotic encryption | $DH_1$ Chaotic interleaving |
| 0 | 16.1 | 19.9 | 15.8 | 15.8 | 15.8 | 15.9 |
| 5 | 37.8 | 75 | 26.8 | 26.9 | 26.7 | 27.0 |
| 10 | 85 | 120 | 66.1 | 71.9 | 66.5 | 54.1 |

**Table 6.2** PSNR of the image transmission over a fading channel

| SNR (dB) | PSNR | | | | | |
|---|---|---|---|---|---|---|
| | $DM_1$ Standard | $DM_1$ BCH (15, 7) | $DH_1$ Non-interleaving | $DH_1$ Block interleaving | $DH_1$ Chaotic encryption | $DH_1$ Chaotic interleaving |
| 5 | 16.8 | 17.9 | 16.7 | 16.6 | 16.5 | 17.2 |
| 15 | 26.7 | 28.4 | 26.4 | 27.6 | 25.2 | 26.2 |
| 25 | 35.4 | 36.7 | 35 | 42 | 37.13 | 43 |

## 6.6 Performance and Complexity

In this section, the FEC complexity and the need of powerful data randomization relation are discussed. Almost, the more complex error control scheme is required to implement for the bad condition communications channel and ensuring the loss of data within the accepted level. The efficient data randomization techniques may increase the required memory, but it reduces the need of complex error control schemes. Finally, it may decrease the computational complexity of the combined technique 'lower complexity FEC scheme and the interleaving technique.' The computational complexity of any process is defined as the number of all arithmetic and logic operations which are required for this process in addition the required memory also. The different WPANs networks are presented to achieve rapid and quick of simple and active wireless communications. Hence, it utilized in its standard weak FEC or not such as Zigbee networks. It is presented for near and direct to communicate different things in specific small areas with limited required power.

In the following, the complexities of the different error control schemes are discussed. In the coding theory, R-S code is an example of maximum distance separation (MDS) codes and it achieves the Singleton bound, which means the Reed–Solomon codes provide reliable reconstruction of the original message. When the channel conditions are known and static, the Reed–Solomon may provide optimal recovering the erasure data if the suitable code rate is utilized to design the R–S encoder. On other hand, there is a vital need for retransmission mechanism, which is still required with the higher data loss, and the original message cannot be recovered. Then, in this situation, the code rate should be modified according to the changes in the packet erasure rates.

The Reed–Solomon is also impractical for large amount of k message symbols. The computational complexity of the decoding process is $O[k(n-k)\cdot\log(n)]$. It is very expensive for the multi-cast and broadcasting scenarios. The mobile terminals in the wireless communications have limited resources, such as computational power resources. Therefore, the lower complexity of the FEC is prerequisites for utilizing in the mobile wireless applications and wide deployment in different applications such as wireless sensor networks (WSNs) [20].

## References


1. Howitt I (2001) WLAN and WPAN coexistence in UL band. IEEE Trans Vehicular Technol, 50(4)
2. Viterbi AJ (1971) Convolutional codes and their performance in communication systems. IEEE Trans Commun Technol 19(5)
3. Howitt I (2001) WLAN and WPAN Coexistence in UL Band. IEEE Trans Vehicular Technology 50(4)
4. Viterbi AJ (1971) Convolutional codes and their performance in communication systems. IEEE Trans Commun Technol 19:5
5. Heller JA (1968) Short constraint length convolutional codes, Jet Propulsion Lab, California Inst Technol, space programs Summary 37–54, vol III,, pp 171–177
6. Omura JK (1969) On the Viterbi decoding algorithm. IEEE Trans Inform Theory IT-15
7. Soliman NF, Albagory Y, Elbendary MA, Al-Hanafy W, El-Rabaie E-SM, Alshebeili SA, El-Samie FEA (2014) Chaotic interleaving for robust image transmission with LDPC coded OFDM. Wireless personal communications, Springer, 79(3):2141–2154
8. El-Bendary MAM, Abou-El-Azm AE, El-Fishawy NA, Shawki F, Abd El-Samie FE, El-Tokhy MAR, Kazemian HB (2012) Performance of the audio signals transmission over wireless networks with the channel interleaving considerations. EURASIP J Audio, Speech, and Music Processing, Springer, 1(4)
9. El-Bendary MA, El-Azm AA (2011) An efficient chaotic interleaver for image transmission over IEEE 802.15. 4 Zigbee network. J Telecommunications Inf Technol, 2011, pp 67–73
10. Mohamed M, El-Azm AA, El-Fishwy N, El-Tokhy M, El-Samie FA, Shawki F (2008) Bluetooth performance improvement with existing convolutional codes over AWGN channel. Proceedings 2nd international conference on electrical engin. design & technologies ICEEDT'08
11. El-Bendary MA (2017) FEC merged with double security approach based on encrypted image steganography for different purpose in the presence of noise and different attacks. Multimedia Tools and Applications
12. Nassar SS, Ayad NM, Kelash HM, El-Sayed HS, El-Bendary MA, El-Samie FEA, Faragallah OS (2016) Content verification of encrypted images transmitted over wireless AWGN channels. Wireless Personal Commun 88(3):479–491
13. El-Bendary MAM, El-Azm AEA, El-Fishawy NA, Al-Hosarey FSM, Eltokhy MA, El-Samie FEA, Kazemian HB (2012) JPEG image transmission over mobile network with an efficient channel coding and interleaving. Int J Electron 99(11):1497–1518
14. Abouelfadl AA, El-Bendary MAM, Shawki F (2014) Enhancing transmission over wireless image sensor networks based on ZigBee network. Life Sci J 11(8):342–354
15. El-Bendary MA, Abou-El-Azm AE, El-Fishawy NA, Shawki F, El-Tokhy M, El-Samie FA, Kazemian HB (2013) Image transmission over mobile Bluetooth networks with enhanced data rate packets and chaotic interleaving. Wireless Networks 19(4):517–532
16. Nassar SS, Ayad NM, Kelash HM, El-Sayed HS, El-Bendary MA (2016) Secure wireless image communication using LSB steganography and chaotic baker ciphering. Wireless Personal Commun 91(3):1023–1049
17. Nassar SS, Ayad NM, Kelash HM, El-Sayed HS, El-Bendary MA (2016) Efficient audio integrity verification algorithm using discrete cosine transform. Int J Speech Technol 19(1):1–8
18. Eldokany I, El-Rabaie ESM, Elhalafawy SM, Eldin MAZ, Shahieen MH, El-Bendary MAM (2015) Efficient transmission of encrypted images with OFDM in the presence of carrier frequency offset. Wireless Personal Commun 84(1):475–521
19. Fridrich J (1998) Symmetric ciphers based on two-dimensional chaotic maps. Int J Bifurcation Chaos 8(6):1259–1284

20. El-Gohary NM, El-Bendary MAM, El-Samie FEA Foud MM, Utilization of Raptor Codes for OFDM-System Performance Enhancing. Wireless Personal Communications. Springer, https://doi.org/10.1007/s11277–017-4248-6
21. Kasban H, Mohsen AM, El-Bendary (2016) Performance Improvement of Digital Image Transmission over Mobile WiMAX Networks,Wireless Personal Communications, Springer

Printed in the United States
By Bookmasters